NEW TECHNOLOGY AND THE PROCESS OF LABOUR REGULATION

NEW TECHNOLOGY
and the
PROCESS OF
LABOUR REGULATION

ERIC BATSTONE · STEPHEN GOURLAY
HUGO LEVIE · ROY MOORE

CLARENDON PRESS · OXFORD
1987

Oxford University Press, Walton Street, Oxford OX2 6DP

Oxford New York Toronto
Delhi Bombay Calcutta Madras Karachi
Petaling Jaya Singapore Hong Kong Tokyo
Nairobi Dar es Salaam Cape Town
Melbourne Auckland

and associated companies in
Beirut Berlin Ibadan Nicosia

Oxford is a trade mark of Oxford University Press

Published in the United States
by Oxford University Press, New York

British Library Cataloguing in Publication Data
New technology and the process of labour
regulation.
1. Labour supply—Great Britain—Effect
of technological innovations on
I. Batstone, Eric
331.12'5'0941   HD6331.2.G7
ISBN 0-19-827274-X

Library of Congress Cataloging in Publication Data
New technology and the process of labour regulation.
Includes bibliographies.
1. Industrial relations—Great Britain—Effect of
technological innovations on—Case studies.
I. Batstone, Eric.
HD6331.2.G7N48 1987 331'.0941 87-11146
ISBN 0-19-827274-X

Set by Hope Services, Abingdon
Printed in Great Britain
at the University Printing House, Oxford
by David Stanford
Printer to the University

PREFACE

THIS monograph reports the findings of four case-studies undertaken in 1983 and 1984. The research would not have been possible without the very great help, advice, and co-operation which we received from union officials, managers, and workers in the four firms. To them we express our gratitude and thanks.

The research was funded by a grant from the Economic and Social Research Council. Much of the preliminary analysis of the data we collected was undertaken by our research secretary, Toni Batstone, to whom we are indebted. Finally, we should express our debt to our colleagues and students, who have provided a stimulating atmosphere in which to work.

ERIC BATSTONE
STEPHEN GOURLAY

CONTENTS

TABLES AND FIGURE

ABBREVIATIONS

APEX	Association of Professional, Executive, Clerical and Computer Staff
ASTMS	Association of Scientific, Technical and Managerial Staffs
AUEW	Amalgamated Union of Engineering Workers
BIFU	Banking, Insurance and Finance Union
EETPU	Electrical, Electronic, Telecommunication and Plumbing Union
JNC	Joint Negotiating Committee
NUBE	National Union of Bank Employees
NUSMWCHDE	National Union of Sheet Metal Workers, Coppersmiths, Heating and Domestic Engineers
TASS	Technical, Administrative and Supervisory Section (of AUEW at the time of the research)
TGWU	Transport and General Workers' Union
TUAC	Trades Union Advisory Committee
UCATT	Union of Construction, Allied Trades and Technicians
USDAW	Union of Shop, Distributive and Allied Workers

I

INTRODUCTION

THE content and control of work has been a major theme within sociology and related disciplines for many years. Much of the debate on the nature of industrialization has centred upon the question of changes in the nature of work and their implications for the broader social condition. Many writers have argued that capitalist development has involved, and indeed depended upon, the progressive reduction of worker skills and an intensification of effort. Others have argued the reverse. Not only have skill requirements increased, it is claimed, but workers' rights within the workplace have also come to receive fuller recognition.

Similar arguments are to be found when there are dramatic changes in the nature of technology, management techniques, or the structure of the economy. For example, from the late 1950s the implications of automation for job content became an important subject of debate. Many saw it as simply accelerating long-term trends towards deskilling or an upgrading of skills. For others, automation reversed past trends towards the degradation of work. In recent years the advent of 'new technology' has once more awakened this long-standing debate.

This book seeks to contribute to this discussion of the effects of new technology upon work through four detailed case-studies. In these we not only compare work organization before and after the introduction of new technology, but also look in some detail at the way in which the changes were brought about. Before outlining our findings, however, we need to look at the general arguments which are of relevance to our research. In this chapter, therefore, we first consider the way in which the question of technology has been approached in previous work and how job content has been analysed. In addition, we discuss the question of the strategies pursued by both management and workers in relation to new technology.

TECHNOLOGY

A fundamental problem concerns what exactly is meant by technology. It has been common to stress that the notion should encapsulate not merely physical pieces of production machinery but also 'social things as well—organizations and processes concerned with human ends' (Nisbet 1971: 41). Hill similarly argues that the term 'embraces the physical organization of production, the way in which the hardware has been laid out in the factory or other place of work', and, he claims, technology therefore 'implies the division of labour and work organization which is built into or required for efficient operation by the productive technique' (1981: 86).

In the most general of terms, such arguments are perfectly valid. But we do not believe that they are very useful in seeking to assess the impact of technology upon work content and control. For, if we follow these definitions strictly, then our dependent variables are included within our independent variable. It is therefore perfectly correct to conclude that job content is determined by technology—this is true by definition. But the conclusion is neither helpful nor illuminating. For our purposes, then, it is more useful to define technology more tightly, if less sociologically, simply in terms of physical pieces of machinery. In short, we need to maintain an analytical distinction between the 'instruments of production' and 'relationships of production'. In practice, even those who espouse a wider approach frequently operate with this more restricted definition.

Classificatory and descriptive variations and weaknesses have in the past also served to confuse much of the discussion of the impact of technology (see e.g. Davis and Taylor 1976: 390–1). Certainly the classification of technology is a complex task, and even writers whose approaches have been widely emulated have frequently noted the limitations of the approaches they have adopted. Joan Woodward *et al.*, for example, appear finally to have despaired of developing a satisfactory classification (1970); Blauner similarly notes a variety of problems (1964:13). The basic inadequacies of classification and description are exacerbated by the fact that categorizations vary according to the focus of the research and differences of view as to what exactly technology consists of.

The basic interests of writers on technology have varied widely. Of central importance for present purposes is the distinction between those who have been concerned with its effects on organizational structure and those who have concentrated upon work experience. Woodward is an eminent example of the former, and her classification was therefore guided by the sorts of problems which different systems of production raised for management. Meissner, an example of the latter interest, sought to develop categories which focused upon the man–machine interface. As Meissner shows, Woodward's classification does not fit well with a classification of man–machine systems (1969: 246).

The problems which arise if categories are simply shifted from one area to another can be seen in the work of Blauner. His concern is with the nature of work experience, but his classification of technology is similar to Woodward's. He focuses upon what he sees as 'characteristic forms of production' (1964: 7) in different industries and assumes that these will indicate the nature of the relationship between worker and technology. This may be valid in some cases, but often such an assumption is misleading and inaccurate. For example, Blauner characterizes the car industry as involving assembly-line production. But the majority of workers in the car industry do not in fact work on assembly lines. Similarly, many workers in the chemical industry—involving continuous-process production—are not control room operators. Most companies employ a wide variety of different kinds of technology. It may be that the problems found in different types of production system do affect individual work experience, by shaping management structure and behaviour (see below), but this is not the primary thrust of Blauner's argument. In short, we need to distinguish between the overall or dominant technology employed in a production system and the nature of the worker–machine interface. Both may be important in understanding the nature of work experience. In some cases the two may be interrelated. However, neither of these points can simply be assumed to apply.

But even if the basic focus of interest is made clear, classifications frequently encounter problems. Important among these is that the way in which a particular classificatory system is used appears to vary considerably between writers, making comparison extremely difficult and secondary analysis potentially extremely misleading (Davis and Taylor 1976: 391). It is for this reason that either clear definitions of categories or descriptive accounts of the technology are of central importance.

In looking at the impact of new technology these problems are exacerbated further because we need to know about both the nature of the basic production system and the characteristics of the new technology which is introduced into it. While some earlier work failed to provide satisfactory accounts in these respects, there has been a growing recognition that, for example, the impact of a particular type of new equipment upon work experience will depend upon the nature of the production system and, very often, upon the precise characteristics of the new technology itself. As will be seen in Chapter 4, for example, variations in the details of different generations and types of CNC (computer numerical control) machine are significant in affecting work experience. Accordingly, in subsequent chapters we outline the general nature of the production system and the character of the new technology, in both cases focusing upon how they relate to the tasks which workers have to perform.

Technological determinism

There have been a wide range of views on the extent to which technology determines organizational arrangements and work experience. Writers such as Kerr *et al.* (1973), Bell (1974), and others have argued that technology has a major impact upon many features of society. Others have claimed that technology is a key factor shaping work experience (Blauner 1964), organizational arrangements (Woodward 1965), industrial action and bargaining (Sayles 1958; Kuhn 1961), and more general orientations and class consciousness (Mallet 1975; Touraine 1971). Much of the earlier work on new technology, particularly that adopting what Wilkinson calls the 'innovative approach' (1983), adopted a similarly deterministic approach (e.g. Evans 1979; Barron and Curnow 1979). However, technological determinism has been subjected to a growing amount of criticism.

First, it has been argued that the relationship between technology and various aspects of social organization both within and outside the factory gates has been grossly exaggerated. Hickson *et al.*, for example, found that technology was far less important in shaping organizational structure than Woodward initially suggested (1969). Gallie (1978) and others have produced evidence which suggests that technology has little effect upon worker attitudes. There has been a growing recognition that work experience is not to be primarily explained by technology (for a review of much of the literature see Davis and Taylor 1976).

However, in their detailed arguments the differences are often rather less great than the general arguments presented might suggest. On the one hand, those who are critical of technological determinism do not wish to deny that technology has some impact. A number of studies have noted, for example, that its effects are greater on those social arrangements which are closest to it (e.g. Davis and Taylor 1976: 411; Pugh and Hickson 1976: 151–4). On the other hand, those who have been labelled as technological determinists have recognized that the significance of technology varies across different elements of society (e.g. Kerr *et al.* 1973: 291–3; it is worth noting that in his more recent work Kerr (1983) has placed even greater emphasis upon this point). They have also recognized the importance of other factors. Blauner, for example, notes the importance of the division of labour, economic structure, and 'organizational character' (1964: 5), while Kerr *et al.* stress the importance of the nature of 'industrializing elites' and 'earlier cultural forms' (1973: 293).

A second focus of criticism of theories of technological determinism concerns the explanation of the link between technology and social organization. The basic determinist thesis is very simple—technology requires that certain tasks be done and therefore determines work organization, organizational structures, and industrial attitudes and behaviour (e.g. Woodward 1965: 79). However, the argument often extends significantly beyond this view. Blauner conflates the impact of 'secular developments in technology, division of labour and industrial social structure', and it is unclear how far these other factors are seen as necessarily associated with technology (e.g. 1964: 182). Similar problems are to be found in his more detailed arguments. For example, Blauner argues that five main factors explain why social integration is higher in the continuous-process industries: a finely graded status structure; lower manning levels; team working; the blurring of the division between manual and non-manual workers; and company prosperity, along with low labour costs. However, contrary to his arguments, none of these are necessarily or inevitably attributable to the nature of the technology employed. Similar weaknesses are to be found in the work of Sayles and Kuhn, who appear to believe that status and reward systems are to be explained primarily in technological terms.

Where associations are found between technology and social organization, many critics of technological determinism have argued that the relationships are to be explained by the conceptions of work held by management (e.g. Argyris 1972; Davis and Taylor 1976: 412). Such possibilities are also noted—though rarely developed—by writers who stress the importance of technology. Woodward, for example, notes the possibility that certain types of manager may be selected for key positions under different technological conditions (1965: 79–80) and that in the long term the technology employed results from 'a series of managerial decisions to serve specific markets, to acquire or build plant, to accept certain types of raw materials, and to address the organization to certain production tasks' (1970: 5). Similarly, there is a striking contrast between Kerr *et al.*'s general claims of the importance of technology and their detailed arguments, in which they emphasize the significance of industrializing élites.

The element of choice in the development and application of technology has been stressed by many radical writers in recent years. Braverman, for example, argues that the primary means by which the capitalist accumulates capital is through the degradation of workers, subjecting them to closer control and removing from them all conceptual aspects of work. Such priorities shape the design of new equipment and the way in which it is employed (1974). Technology is therefore associated with particular sets of work experience within capitalism, but these should be attributed to the strategy of the capitalist rather than technology as such. The basic argument of Braverman is open to serious questioning, as will be shown below. The point for present purposes is that he exemplifies one significant form of criticism of technological determinism.

More recent work on new technology, particularly that deriving from case-studies, has increasingly rejected any form of technological determinism. Emphasis has been placed upon more general characteristics of management strategy, organizational structure, and industrial relations in explaining the impact of new technology upon work organization (e.g. Penn and Scattergood 1985; Duhm and Muchenberger 1983; Rose and Jones 1985). In many respects, therefore, the analysis of new technology is incorporated into a more general analysis of the forces which serve to shape the organization of work. It is this approach which we adopt here. However, we seek to develop it somewhat further by looking in considerable detail at the processes by which work organization is shaped with the introduction of new technology, and by highlighting particular characteristics of union organization which shape these processes.

In this section we have stressed a number of points. First, we have argued that in order to assess the impact of technology we need to define it fairly strictly, if less than ideally, in terms of machinery and equipment. Second, we have argued that many discussions have been confused and confusing due to the categorizations of technology employed: we shall avoid such methods and seek to describe the technology in terms of the functions it performs and the nature of the (variable) man–machine relationship. Third, we have pointed to a number of weaknesses in theories of technological determinism (although we have also noted that positions are in reality often less extreme than much of the literature suggests). In particular, such theories have failed to distinguish between necessary, causal relationships and simple empirical associations: they have often failed to consider carefully enough the relative importance of technology as compared with other factors, and they have paid scant attention to questions of technical design and choice.

JOB CONTENT AND LABOUR REGULATION

A second area where some introductory discussion is necessary concerns the approach to be adopted to the question of job content and labour regulation. We use the latter term to refer to the totality of forms of control, reward, and sanction which exist within the workplace. This area has been the subject of a great deal of discussion, stemming from a diversity of interests. Managements have often tried to develop systematic analyses of job content as a means of task allocation and co-ordination, to determine pay and grading, or to build up organizational structures. Within the academic literature, the largest set of studies has probably concerned the impact of job content upon worker attitudes. Such work has concerned not only the question of job satisfaction, but also the measurement of alienation (typically subjectively defined) and class consciousness. Another central concern has been with the distribution of power and control within the workplace. Recently much of this work has been informed by a labour-process perspective, although a great deal of labour history and work within industrial relations has long been concerned with similar issues.

Given this diversity of interest, it is not surprising that the approaches adopted demonstrate a similar diversity. Many studies concern themselves with only particular aspects of the work situation, such as supervisory styles. A great deal of research has relied upon workers' subjective assessments of a limited set of aspects of the work situation—task interest and variety, relationships with workmates and supervision, and so on. Such work typically confronts problems relating not only to the range of factors investigated but also to the actual methods employed. It is clear that findings often vary significantly between structured interviews and less formal discussions. Moreover, the criteria employed by respondents in assessing responsibility, for example, are likely to vary considerably between respondents.

Other studies have used less than adequate data. Instead of looking at the actual work situation, reliance is often placed upon the written claims of managers or specialists. This, as Kusterer notes, is an especially serious weakness in

Braverman's seminal work. He points out that Braverman 'arrived at his conclusions not by studying the activities of actual workers nor even by studying the activities of actual managers, but by studying managerial writings about the increasing control they hoped to acquire . . . Not for the first time, a Marxist has attributed all active power to management, reducing workers to the role of helpless victims of their own increasing immiseration' (1978: 188).

A related problem concerns the lack of balance and perspective in many works. Marginal developments are often endowed with the significance of major transformations. Littler, for example, has claimed that the introduction of the Bedaux system was a decisive step in labour control in Britain (1982). His argument rests upon the fact that at some point in the inter-war period 245 firms used it. This is a minute proportion of all firms, and he provides no evidence that they employed a significant proportion of the labour force. Furthermore, it is clear that the application of the system rarely involved major job redesign, particularly for key groups such as craftsmen, and was frequently manipulated at shop-floor level. Hence, as Zeitlin has pointed out, 'by focussing carefully on the transmission of Taylorist techniques, Littler has paradoxically demonstrated their marginality to British industry' (1983: 371). Similar problems are found in much of the discussion on the rise of scientific management and Taylorism, for example in the work of Braverman. As many later writers have recognized, the significance of these techniques was limited (e.g. Edwards 1979: 97–104; Burawoy 1985; Nelson 1975; Merkle 1980).

The significance of developments which Edwards construed as important new forms of labour control can similarly be questioned. In seeking to show the importance of technical control he draws upon the findings of Yanouzas (1964) to support his contention that supervision becomes less important when machines control workers. But even if we accept his reworking of Yanouzas's data, two problems arise. First, the difference in the time spent by supervisors on labour control under personal and technical control systems is relatively small, and labour control still takes up nearly three-quarters of supervisors' time under the latter system. Second, he claims that the reduction in the frequency of supervisory direction of work is of particular significance. However, under technical control supervisors spend more time evaluating worker performance, and also on discipline. It is not clear why these should be seen as less significant than direction; indeed, in terms of affecting worker attitudes, the reverse might seem equally plausible. In short, the limited differences in the role of supervisors raise questions concerning the utility and significance of Edwards's distinction between personal and technical forms of control.

However, a crucial problem concerns the dimensions of labour regulation which are distinguished. Many writers have sought to identify long-term trends in the development of work, and to this end have focused upon one or two key dimensions, such as trust (Fox 1974), skill (Braverman 1974), the application of scientific management, or the blossoming of formal rules. Given the scope of argument and the problems of data availability, the search for key dimensions is understandable. But such arguments need to justify the choice of dimensions, showing that they encapsulate or shape other important aspects of the work situation. At the other extreme, detailed case-studies are more able to identify the complexities of labour regulation, but all too frequently writers jump too rapidly to 'the grand argument'.

The focus of studies, as we noted above, has varied significantly. Some writers are concerned primarily with the experience of work and the effects it has upon the human potentiality of the worker. Underlying this approach there is frequently a belief that work should be fulfilling in some way (cf. Anthony 1977). This requires that the worker exercise some control over both the tasks performed and his own behaviour. Other studies have concentrated upon the control strategies employed by management. This has been the major thrust of the labour-process literature. Frequently there appears to be a belief that management control is the inverse of worker control—that is, control is zero-sum. But this is not necessarily the case. Indeed, bureaucratization and the fuller specification of rules—often seen as important forms of management control—have provided the basis for worker control of both an individual and collective nature (for

a fuller discussion of this point see Batstone *et al.* 1984).
There are not only problems of who is in control; the question also arises of what is controlled. In much of the labour-process literature there is a risk either that key aspects of work are ignored or that everything is seen as a form of control over labour. Braverman, for example, confines his discussion to skill and the relationship between conception and execution in task performance. No account is taken of the control which workers may obtain through systems of industrial suffrage. But presumably Braverman would not wish to argue that the slave who is a craftsman is less degraded than the semi-skilled freeman.

Edwards (1979) is an example of seeking to include too much within the ambit of management strategies to control labour. He argues that management adopted technical control—where machinery directs and possibly evaluates worker performance—when older systems of control met with resistance. He starts his argument by recognizing that 'how something is produced is in large part dictated, of course, by the nature of the product and by the known and available technologies for producing it'; he also notes the relevance of 'cost considerations' (p. 111). He goes on to point out the 'less well known' point that 'there is also an important social element in the development and choice of technique', namely the control of labour. How less well-known this really is, among workers and academics, is a question which need not detain us here. What is of greater importance for our purposes is that his subsequent discussion focuses exclusively upon the development of technology as a form of labour control, so that his analysis is remarkably one-sided. More than this, it thereby ignores potentially crucial issues concerning the relationship between other factors, such as the nature of markets, technical production problems, relative factor costs etc., and labour control. Such distinctions—particularly that between production and labour controls—will form an important theme in the following chapters.

What is striking about the literature on job content and control in the work situation, then, is the lack of careful and systematic studies. Blackburn and Mann, for example, noted—with some exaggeration—that their own work and that of Turner and Lawrence were the only systematic attempts 'to measure jobs in the sociological literature' (1979: 46). Things have changed little over the last few years. In an attempt to identify the key aspects which need to be covered in a satisfactory analysis, we shall look first at what has been the exclusive focus of many studies—task content. Secondly, we shall consider the broader aspects of the work situation which are of relevance and which are included in our notion of labour regulation. But in addition to identifying the key dimensions of labour regulation, we also need to consider the forms which such regulation take; this will be the third theme in the following discussion.

Task content

A central theme in many discussions of task content has been skill. In one of the most influential studies of the last decade or so, Braverman made it a central concept in his argument that the capitalist increases his control through the degradation of labour (1974). We have already noted the limitations that this approach imposes. Here we are concerned with the adequacy of Braverman's conception of skill.

He defines skill largely by reference to an ideal-type craftsman: 'For the worker, the concept of skill is traditionally bound up with craft mastery—that is to say, the combination of knowledge of materials and processes with the practised manual dexterities required to carry on a specific branch of production' (1974: 443). The thrust of his argument is that the former has been lost to 'scientific, technical and engineering knowledge', which has been monopolized by management. There are a number of problems with this argument.

If we confine our discussion, as Braverman does, to a notion of craft, then clearly a major proportion of the labour force both now and in earlier phases of industrialization fits uneasily within the analysis. But there are also other difficulties which are of greater importance for our purposes. First, Braverman's conception of craft work is highly romanticized. Such an approach is common. Wright Mills, for example, talks of craft work involving 'no ulterior motive than the product being made and the processes of its creation'. For the craftsman, he claims, 'there is no split between work and

play, or work and culture. The craftsman's very livelihood
determines and infuses his entire mode of living' (1951: 220).
If we dig under the noble sentiments which such statements
embody, we come across some interesting problems and
issues. For example, the print craftsman 'touching up'
pornographic photographs—a highly skilled task—identifies
strongly with his product, while his counterpart in an
armaments factory derives meaning from the purposes to
which his product will be put. The craftsman, apparently, will
be little concerned with the welfare of his dependants and his
wage packet and, as long as the employer wants high-quality
work, he will have few problems of labour control. This
picture scarcely conforms to reality.

In addition, it is difficult to differentiate between the
knowledge involved in craft mastery and 'scientific, technical
and engineering knowledge'. Indeed, in many cases some
basic knowledge of the latter kind is essential for competent
performance on the part of the craftsman. Nor is it clear why,
or how, management should monopolize new forms of
knowledge. Indeed, in areas such as the maintenance of new
technology it would seem that managements are often keen
that craftsmen acquire such knowledge and that it is
craftsmen who are sometimes resisting such moves.

Braverman also avoids many of the central problems
associated with the analysis of job content—those concerned
with weighting the different dimensions identified. He claims
that craft knowledge declines, with no compensatory increase
in scientific knowledge, while manual dexterity appears to
remain constant or fall. But if we reject this argument—and
the reasons for doing so have been noted in the preceding
paragraph—then we have to be able to assess the degree and
significance of movements along these various dimensions, if
we are to make general statements concerning overall trends
in skill. If only crudely, we have to be able to say that there
was a large fall in craft knowledge, for example, but only a
small increase in scientific knowledge. Even then, we might
conclude that there was an overall increase in skill, if we
attached much greater significance to the scientific-knowledge
dimension than the craft dimension.

This problem, of course, is not peculiar to Braverman. It is most evident in the work of writers, such as Blackburn and Mann (1979), who seek to build up summary indicators from scores on a multiplicity of dimensions. More generally, the utility of such summary measures is open to question, given our present state of knowledge. Since our interest is the effects of new technology, it seems more useful not to move to such summary scores and, instead, to highlight the way in which jobs move in different directions on different dimensions. This is the approach which we shall adopt in Chapter 6, when we seek to summarize our findings.

A further difficulty with the concept of skill, and indeed other measures of task content, concerns variations in skill requirements between tasks and situations. In many jobs little skill is required for much of the time but particular situations, notably crises, may require very high levels of skill. The assessment of the skill involved in a job is therefore likely to vary depending upon, for example, whether we weight skill requirements according to the amount of time they are actually employed, or place greater emphasis upon those situations requiring a higher degree of skill. While this problem is rarely explicitly discussed, it seems that the latter approach is generally adopted. We shall also follow this convention.

As many commentators have recognized, the notion of skill is particularly difficult to operationalize. The fact that 'skill is measured less by a formal definition than by historical context and comparison' (Thompson 1983: 92) does not escape this problem. Some writers have attempted to find proxies for, or indicators of, skill. Formal gradings and job titles are of little use if one is trying to measure the objective content of jobs. Indeed, while he fails to assess the precise significance of his argument, Braverman's emphasis upon the dubious classification of occupations is quite correct. As many writers have noted, the labels 'craft' and 'skilled' are distributed through social processes. Often the term 'craft' reflects the successful pursuit of particular strategies on the part of a work group, rather than any inherent features of the job (Turner 1962).

Four other approaches which have been widely used to assess task content and worker autonomy concern training times, the detail of rules applying to work performance, the range of tasks undertaken, and the 'time-span' of discretion. All of these approaches encounter problems. The duration of training is a classic issue of negotiation and varies widely. Strictly we should have to conclude that, if two people were performing the same job and only one of them had been formally trained, then the other was less skilled. Given wide variations between countries in the time committed to training, we should also be forced to conclude that many comparable jobs were less skilled in Britain than, say, in Germany. To operate on notional training times—how long it would take for someone to become competent in a job—leaves a great deal of scope for subjective assessments.

The amount and detail of rules relating to task performance have also been widely applied as an indication of the degree of control exerted over task performance (e.g. Crompton and Jones, 1984: 59). Where there are few rules, it is suggested, there is greater scope for worker autonomy and the exercise of skill or discretion. A few examples indicate the problems of such an approach. Many jobs which would generally be deemed unskilled are shaped by remarkably few rules (e.g. the jobs of cleaners), while many jobs which are defined as highly skilled—those of airline pilots, for example—are characterized by a multiplicity of detailed rules and regulations. What such approaches fail to do, among other things, is to take account of the significance of the task undertaken. It may not matter if a cleaner fails to polish a table, but the consequences of a pilot failing to put the wheels down on landing would.

In large part influenced by the debate concerning trends in the division of labour, another approach to the analysis of job content has concerned the range of tasks which a worker performs. Again, this encounters difficulties. As many critiques of job enlargement have noted, a greater range of tasks need not mean greater creative opportunities or influence for a worker. Nor need they mean greater autonomy, as debates over greater job flexibility indicate. Not only does the significance of task range depend upon the nature of the tasks, but it also depends upon the extent to which workers can choose when, how, and if they undertake particular tasks.

The time-span of the discretion approach to the analysis of

jobs focuses upon the period of time for which 'it would be possible to exercise inadequate discretion before that fact would come to the attention of the immediate manager' (Jaques 1967: 96). The greater the time period, the greater the discretion. The weaknesses of this approach are similar to the rules approach. It fails to take sufficient account of the responsibility and significance of tasks, and also their immediacy. Many jobs—particularly peripatetic ones such as those of street cleaners and postmen—have a long time-span of discretion simply because it would be costly to supervise them in detail. In many other jobs, which in fact involve very significant degrees of discretion, errors would very quickly come to light. Amongst such groups might be surgeons, those controlling dangerous production processes, or those monitoring and controlling the supply of electricity. The time-span of discretion is likely to vary significantly according to the precise nature of the task and the nature of the error.

Worker control and discretion may be as important in a negative, as in a positive, direction. Much of the analysis of job content has concentrated upon the positive, or creative, aspects of work. Dubois *et al.* (1976), however, have pointed to the importance of 'sins of omission': that is, the extent to which workers may affect production through the failure to act according to the rules. Moreover, in looking at both positive and negative forms of control, they note a variety of areas of potential worker influence—output, quality, working methods, and cycle times. A related distinction has been highlighted in much of the discussion of the demands which automation places upon workers. For it has been argued that under automated systems the traditional link between worker effort and output is broken. What then becomes important is a set of skills associated with monitoring the production process. These may involve not only attentiveness but also the ability, based upon a detailed practical understanding of the production process, to identify errors and react swiftly and effectively in the case of breakdowns. Hirshhorn has termed this 'second order control' (1984).

We have cast doubt upon the utility of many single-factor approaches to the analysis of work. More satisfactory are the more detailed and systematic analyses undertaken by such writers as Turner and Lawrence (1965), Blackburn and Mann (1979), and Dubois *et al.* (1976). However, these also encounter a number of problems. That concerning the summation of scores on different dimensions has already been touched upon. In addition, the focus of their work varies. The work of Turner and Lawrence, and to a lesser degree of Blackburn and Mann, focuses upon what claims a job makes upon the worker in terms of mental demands, effort levels, and so on. Dubois *et al.* are concerned less with such factors than the degree of influence which workers exert over the production process (as well as the degree of worker independence, which we will turn to below). In our view, both aspects are significant.

Accordingly, when we come to summarize the effects of new technology upon task content in Chapter 6, we shall follow Dubois *et al.*'s approach concerning the aspects of the production process which workers may affect. But in addition we shall consider two other features of task content. The first concerns the demands which competent performance requires of the worker, in terms of such elements as effort levels, responsibility, and task range. The second concerns the various forms of controls and sanctions under which the worker operates; these are discussed more fully below.

Broader aspects of labour regulation

Much of the literature on work has confined itself to a consideration of task content. But work experience involves far more than the performance of certain production or service functions. Many workers spend a significant proportion of their waking hours at work. During that time they are not simply or solely concerned with production tasks. Indeed, some studies have suggested that the proportion of work time which is actually spent in performing work tasks may be very low. The precise reasons for this need not detain us here. The important point is that work is a social situation. A failure to recognize the importance of this point has marred a great deal of industrial sociology. For example, while considerable stress has been placed upon the way in which patterns of social interaction outside of the workplace may affect attitudes to work, there has been insufficient attention paid to similar social pressures within the workplace (see e.g. Batstone *et al.* 1975).

Once we define the work situation in these broader terms, and extend our focus beyond the more immediate aspects of task performance, a number of factors appear to be of particular significance in developing an adequate analysis of work experience. Central among these is the notion of industrial suffrage. Many writers, particularly within the liberal pluralist tradition, have stressed the importance of the development of a variety of worker rights within the work-place in the course of industrialization. Workers may achieve a significant degree of control over their own life chances within the place of work through legislation, collective bargaining, individualistic strategies, or employer initiatives. Other writers have tended to belittle the significance of such developments, arguing that their main significance lies less in the way in which the lot of the typical worker is improved than in the way in which such systems serve to integrate workers into the very system which exploits them, thereby diverting workers from challenging that system. It is from this perspective that many labour-process writers have touched upon broader patterns of labour regulation. They stress the way in which these are developed to strengthen the legitimacy of management in the eyes of workers. Nevertheless workers may achieve some influence over their day-to-day experiences within the workplace through these innovations. In this area, therefore, we turn to issues which have classically been the concern of industrial relations.

Rarely have systematic analyses been undertaken of these more general areas of labour regulation, despite the fact that they are central to a wide range of debates. Only a limited number of surveys have looked at such matters, and rarely in a manner which permits a coherent picture. The main systematic studies of this kind are relatively old. They include Goodrich (1920), Hilton *et al.* (1935), and Slichter, Healy, and Livernash (1960). These works, however, concentrate upon only one, albeit important, aspect—the extent of joint regulation and union influence. Nevertheless, in doing so they do high-light a range of areas and issues of more general importance. These can usefully be discussed under four main headings.

The first of these concerns the rules and controls which serve to shape task performance. Hence, for example, unions or workers may play some role in the design of jobs and the various standards and rules which relate to work performance. Craft unions, for example, have classically controlled particular job territories and thereby ensured the relative autonomy of the craftsman. Other workers may achieve broadly similar controls by bargaining over job allocations, operating rules, the conduct of supervisors, and effort and output levels. We need, therefore, to look at how far workers influence the task-based controls employed by management.

A second issue is the way in which labour requirements are met by job allocation. This includes a wide variety of factors ranging from recruitment and redundancy to the use of secondary forms of labour, flexibility, and transfers and training.

Third, the more general rewards and conditions of workers within the workplace require consideration, as has been noted by many writers within the dual labour market tradition (e.g. Hodson 1983; Osterman 1984). Of most obvious importance here are questions relating to pay, hours of work, and other conditions. Also relevant are many issues which have been the subject of considerable discussion among later generations of labour-process writers and typically discussed as aspects of the internal state or bureaucratic control (e.g. Edwards 1979; Burawoy 1985). These relate to endeavours by the employer to shape worker attitudes, as well as constitutional provisions for the handling of grievances. Many of these factors concern the independence of workers and the extent to which management is able to impose controls upon them, both in relation to task performance and more generally (Dubois *et al.* 1976).

A fourth area of considerable significance is the extent to which workers influence the more general strategies of management which shape the context in which work occurs. This area has been the centre of discussions concerning industrial democracy in recent years and, of course, it is here that workers have classically enjoyed least influence. But it needs to be incorporated into any thorough analysis of the pattern of labour regulation.

In this section, then, we have argued that there is a need to look at these wider areas of labour regulation and also to investigate the extent to which workers achieve some influence both in and through them. Many labour-process writers have paid scant attention to the significance of the latter point. Workers may achieve considerable influence and control over management through involvement in the creation and application of rules—they may mobilize institutional arrangements in their favour. Rules also provide a gauge for the actions of management, as well as providing a form of worker defence against changes in management demands. Furthermore, they often provide a fruitful area of continual negotiation—rules may be mutually inconsistent, arguments can develop concerning priority of application and, for example, whether the spirit or the letter of rules should be followed (for a fuller discussion of these themes see Batstone *et al.* 1984; Batstone 1984). Such influence may not constitute a fundamental challenge to the structure of capitalism, but it may significantly affect the day-to-day position of workers.

Forms of management and worker control

So far we have been concerned primarily with the areas of labour regulation which need to be incorporated into our analysis. We have also recognized the need to look at the nature of work experience, or the sorts of preconditions of competent performance on the part of the worker. We have just noted the importance of one aspect of control forms, namely the degree of joint regulation. But a great deal of recent discussion has centred upon the more general forms of control which management employs. Although the distinction cannot be pursued fully, it is convenient for our present purposes to identify two main approaches to this question. The first of these focuses upon the underlying rationale or philosophy of control; the second concentrates upon more concrete factors.

A wide range of studies differentiate between the rationales of control systems. Etzioni, for example, seeks to relate different forms of control strategy on the part of employers to different sorts of motivation on the part of employees (1961). He distinguishes between three resulting types of compliance structure—coercive (based on force), utilitarian (where financial rewards are dominant), and normative (where ideological factors play an important role). However, the most common approaches have concerned the extent to which management seeks to achieve its goals through worker commitment rather than through detailed control. This is the central distinction, for example, in the work of McGregor (1960) and many writers within the human relations and neo-human relations schools of thought.

More radical writers have also pursued similar themes. Fox, for example, stresses the importance of trust and distrust as 'fundamental principles implicitly informing the ways in which men organize, regulate and reward themselves for the production of goods and services, and the significance of these principles for their wider social relations' (1974: 13). He distinguishes between two ideal types of high and low discretion. The 'fundamental and basic' characteristic of the low-discretion syndrome is that 'the role occupant perceives superordinates as behaving as if they believe he cannot be trusted, of his own volition, to deliver a work performance which fully accords with the goals they wish to see pursued or the values they wish to see observed' (1974: 26). He then goes on to specify a number of other characteristics of the low-discretion syndrome. Important to his argument, however, is that—as with Etzioni and many other writers—it is assumed that a relationship exists between the rationales underlying management's behaviour and the attitudes which workers develop towards the employer and their work.

Within the labour-process tradition, Friedman adopts a similar approach. He identifies two broad strategies which management pursues. The first largely conforms to the Braverman model of close control. The second, responsible autonomy, 'attempts to harness the adaptability of labour power by giving workers leeway and encouraging them to adapt to changing situations in a manner beneficial to the firm. To do this top managers give workers status, authority and responsibility' (1977: 78).

Other writers have placed rather greater emphasis upon more concrete forms of control, albeit seeing these as embodying particular rationales. One approach of this kind which has received a good deal of attention is by Edwards

(1979). He identifies three broad types of control which, he claims, have tended to develop sequentially over time. Changes have been prompted by worker resistance to existing controls and have led to a progressive institutionalization or structuring of control, reducing the extent to which authority can be seen by workers as taking a personal form. This is deemed to be important as a means not only of securing surplus value, but also of obscuring that fact. The earliest form of control was by direct, personal supervision, where the key elements of control—identified as direction, evaluation, and discipline—were all at the personal discretion of the manager or supervisor. Subsequently, Edwards argues, managers sought to develop more structural forms of control by building it into machines which then directed, and possibly even evaluated, worker performance. A third form of control was then introduced—bureaucratic control—informed by the 'principle of embedding control in the social structure or the social relations of the workplace' (1979: 21).

The attraction of these approaches clearly lies in the way in which they seek to integrate a diversity of variables into a limited number of categories. But this can also be their weakness. To be of use, such classifications need to meet three requirements: first, that they identify what are major differences in forms of control; second, that they combine like with like; and third, that they are also capable of differentiating like from unlike. In practice they often fail to do so. On the first point, for example, the significance of some of Edwards's distinctions has already been questioned.

Classifications frequently employ ideal types. These, by definition, seldom correspond simply with empirical reality. Real situations are located somewhere between the extremes of the ideal types and rarely is guidance given as to how differentiations should be made between actual control systems. This is particularly true of approaches such as that of Fox, as he notes at various points (e.g. 1974: 120). Similar problems arise in the work of Edwards. Not only is there reason to doubt the historical accuracy of his model (see Van Houten 1980; Batstone *et al.* 1984), but in addition it is unclear what happens, for example, to technical control when bureaucratic control is introduced. The two forms clearly embody quite different principles, and this might, in itself, be expected to lead to tensions. In fact, it is possible to argue that by underplaying the mix of rationales within control systems these approaches ignore one of the most important and interesting aspects of the control process.

Indeed, it is common for writers to suggest that there will be a variety of pressures to eradicate such inconsistencies. Both Etzioni and Fox, for example, assume that the type of control system employed by management will be mirrored in the attitudes and behaviour of workers. Etzioni argues that this will come about through the existence of strains and problems of effectiveness in situations of incompatibility. In this way he is able to reduce nine logical categories (the various combinations of three types of control and three forms of compliance) to only three. But there are a variety of reasons to doubt the speed, and indeed significance, of such moves towards an equilibrium position. For example, such adjustments may be made difficult by the very nature of the incongruency, the risks involved in shifting strategies, the occurrence of shocks and crises during any movement, and the time period involved in the move to equilibrium.

One writer who does recognize the tensions involved in control strategies in a systematic manner (rather than producing them as a *deus ex machina* to explain changes in control strategies, as Edwards does) is Friedman (1977). He recognizes that, under certain conditions, both direct control and responsible autonomy strategies may lead to problems for management and, consequently, pressures for change. In other words, far from there being a stable equilibrium, there may be important forces moving employers away from what some writers would define as stable situations. One important factor here has been touched upon previously—namely, the way in which workers may exploit control systems. This may occur directly when workers, for example, use the rules as a form of defence. Or it may occur more indirectly as attempts at management control lead to new forms of worker control in other areas (this appeared to be common, for example, where employers changed from piecework to measured daywork).

The question of how particular forms of control actually work rarely receives the attention it deserves. For example, considerable emphasis has been placed upon the way in which machines may pace workers. But the significance of this depends upon a wide variety of factors—the actual speed of an assembly line, the range of tasks a worker is expected to perform, the ability to move up and down the line, the extent to which workers are able to reorganize work amongst themselves, the frequency of breakdowns, and so on (see e.g. Batstone *et al.* 1977). In addition, even if the line presents workers with particular tasks, there is no guarantee that they will be adequately performed, or indeed performed at all. That is, a further set of questions concerning why workers accept the demands of the machine have to be answered. Similar points apply to bureaucratic control structures, where frustrated promotion expectations, failure to provide job security, inter-grade rivalries, and frustrated expectations which these systems create may all exacerbate problems of control and of the legitimacy of the employer in the eyes of the workforce. In short, as many managers are only too fully aware, organization and control systems involve the art of the second-best, and may therefore require periodic adjustments.

A related problem concerns the extent to which, for example, a high discretion or high trust situation—in terms of actual forms of control—need be informed by a corresponding philosophy. It is common for craftworkers to enjoy a high degree of discretion. But this is often attributable less to a high trust philosophy on the part of management than to the power of craft unions. In many respects, therefore, the situation may often be characterized by low, rather than high, trust. In other cases, the appearance of high trust may reflect the impracticability of close control, since work is not of a routine nature. Here, then, it is not high trust but the particular contingencies encountered by the firm which explain the pattern of control. Moreover, degrees of 'trust' may vary between various areas of work, while trust, though significant, may not be as great as it might be owing to low trust considerations and fears on the part of management.

Some categorizations are also less than satisfactory, since they include a diversity of control systems within a single category. This is true of Friedman's concept of responsible autonomy. Within this he seeks to include two very different forms of control—systems of joint regulation with powerful trade unions, and individualistic techniques such as job enrichment and semi-autonomous work groups. Even if we accept the significance of the latter, it is widely accepted that these two systems are dramatically different (e.g. Fox 1974). Joint regulation need provide little autonomy to workers individually and rests primarily upon the brute facts of collective worker power, a quite different principle from that involved in more individualistic techniques. It would therefore seem desirable to differentiate between these two systems.

A similar problem arises in the work of Edwards. There is a tendency within his categorizations to shift focus. The primary emphasis within his discussion of personal and technical control appears to be upon task performance, while in discussing bureaucratic control more stress is placed upon the broader system of labour regulation. In part this is understandable, since the logic of the latter is to achieve a more indirect and normative influence upon workers. However, as was implicit in a previous point, different types of control may operate over different areas or aspects of control. It is quite possible, for example, to have personal control which is little regulated by rules within a bureaucratic structure which stresses promotion and incremental merit payments. This would appear, for example, to be the case in many large Japanese companies (see e.g. Batstone 1984).

A growing number of writers within the labour-process tradition and its offshoots have begun to recognize the tensions and contradictions embodied within structures of control (see e.g. Krieger 1983; Edwards and Heery 1985; Knights *et al.* 1985). This development has come about largely because more systematic empirical research has highlighted the very real difficulties in applying these general categories. It has become evident that the attribution of companies to particular categories is arbitrary and that the categories themselves do not encapsulate important differences between overall control systems (see e.g. Deaton 1985, concerning Fox's categorization of management industrial relations strategies, and Batstone and Gourlay 1986). Furthermore, there has been a growing recognition that a variety of factors intrude upon the design of control systems.

A second weakness of these attempts to differentiate between major forms of employer control over labour is that they isolate this issue from other pressures and constraints upon management. This has been increasingly recognized (e.g. Nolan 1983; Kelly 1985). Indeed, the fundamental weakness of much of the earlier labour-process literature was that it was reductionist. Worker control was the central problem for the capitalist. Given this, strategies could be simply 'read off' from this basic and fundamental fact. Of course, many other considerations shape the way in which management achieves profitability, and these, in turn, impose constraints upon the forms of labour regulation which can be adopted.

Edwards, along with many other labour-process writers such as Burawoy, has stressed the importance to the capitalist not merely of securing, but also of obscuring, the acquisition of surplus value. This is deemed to be achieved, it has been noted, by making control more structural. Such arguments raise two questions. The first concerns the importance of the need to obscure the fundamental processes of achieving profit, the second relates to the efficacy of more structured control.

The theoretical centrality of the need to obscure surplus value arises only if one assumes that workers are inherently revolutionary. If this were the case, then it is difficult to believe that more structured forms of control would delude them about the realities of their condition. Indeed, in Edwards's case this problem is all the greater, since workers become aware of exploitation at particular times, only to be deluded once more when new forms of control are introduced. This is difficult to accept—he provides no satisfactory explanation for variations in worker resistance. This brings us to the second question raised above—namely, the efficacy of more structured control. It is not immediately clear why, for example, bureaucratic control should be more effective in obscuring the realities of existence than personal control. Indeed, Edwards himself appears to recognize this in discussing the strains which develop in particular forms of control. But he explains neither precisely why new forms of control delude workers, at least initially, nor why they are subsequently able to remove the blinkers from their eyes. (It might also be noted that many employers seem to be remarkably unaware of this need to obscure the acquisition of surplus value—certainly, it does not appear to figure centrally in debates at board level.)

The one-sidedness of many approaches to labour regulation can also be seen in discussions of internal labour markets. It is often argued that such arrangements are introduced to divide workers and foster some form of normative commitment to the employer on the part of workers. We have already questioned how effective such methods may be. But in addition others have suggested that there are quite different reasons for the establishment of such systems, which are less directly concerned with the control of labour. Important amongst these are the company-specific nature of skills, which results in the employer needing to provide incentives for trained workers to stay with the firm (e.g. Doeringer and Piore 1971). Similarly, it has been argued that attempts to win the normative commitment of workers may demonstrate a similar dependence of management upon workers, particularly under more advanced systems of production (e.g. Offe 1976). These may be seen as forms of control, but they assume a very different significance from that given to them by a great deal of the more radical literature.

The preceding discussion suggests, then, that the simple categorizations of forms of control are less than adequate. There are typically a multiplicity of forms of control within the workplace and, at least at our present state of knowledge, it seems more useful to trace these complexities rather than to ignore them. This is particularly so since a further area of potential significance is then opened up—namely the interaction, and possibly the contradictions, between different forms of control (e.g. Batstone *et al.* 1984). However, we also need to recognize that labour control is not the sole, or even primary, concern of management. Then we can begin to look at the interaction between forms of labour control and management control of the broader production process. It may also help in understanding any lack of coherence in patterns of labour regulation.

THE QUESTION OF STRATEGY

In the preceding discussion we have at a number of points touched upon the inadequacy of the treatment of the question of management goals. It was noted that a major weakness of theories of technological determinism was the failure to recognize the way in which the strategies of key actors might shape the impact of technology. A somewhat different problem arises in relation to questions of labour regulation. The labour-process tradition, at least until recently, has adopted reductionist arguments which treat labour control as the central dilemma of management. Growing concern has been expressed, however, over the fact that there often appears to be little coherence in the policies which management adopts towards labour. Indeed, there are signs of a growing agnosticism among students of labour control—a belief that it is not possible to come to grips with the way in which management treats labour, at least in any simple manner (for a somewhat convoluted discussion along these lines see Storey 1985).

Certainly it would seem that British companies rarely pay much systematic attention to the question of labour regulation. More importantly, perhaps, such considerations rarely figure very centrally, if at all, in the formulation of more general strategies. In part, this appears to reflect the structure and competence of British management. British companies, for example, are more financially orientated than many of their other European counterparts, and this appears to militate against a serious consideration of labour issues (Batstone 1984).

In part as a result of this broader structure, there is a tendency for different aspects of labour regulation, as we have defined it, to be handled by different groups of managers, often with little co-ordination between them. Line managers or industrial engineers, for example, may be concerned with the design of work, while personnel specialists deal with industrial relations and welfare matters. Rose and Jones, for example, have recently pointed to the way in which these two areas tend to be treated separately, with the consequence that changes in work organization do not feed into trans-formations of the broader pattern of industrial relations (1985). However, we would expect that the extent to which this is true, and the precise way in which such patterns can best be interpreted, will vary. Where work organization and industrial relations are integrated, significant changes in work organization will have to be introduced through this system of joint regulation. The latter may therefore impose constraints upon the scale of change in the former.

It is, of course, possible for employers to seek to change the broader pattern of industrial relations. This appears to be most easily achieved where the employer can, as it were, start from scratch—where a new plant is being set up on a green-field site or where new production systems can in some way be isolated from existing production facilities and labour regulation systems. In other cases, employers may find themselves in a position—for example, confronting serious competition, serious challenges from the workforce, or strong political pressures—which induces them to seek major changes in the general pattern of labour regulation.

But where large-scale changes are to be introduced into patterns of labour regulation in which unions play a significant role, management has to work through the system of joint regulation in order to achieve change. The ease with which this can be done will clearly depend upon the way in which unions believe that the proposed changes will affect their members. If there is union opposition, broader changes will be difficult, and change will in any event be constrained by the system of joint regulation. In short, the difficulties of transforming the broader pattern of regulation may serve to discourage management from grand strategies of change (e.g. Kochan *et al.* 1984; Ogden 1982).

Even where management does seek to change both work organization and the broader pattern of labour regulation, they may encounter serious difficulties. They have to work through the old to the new. It may be difficult to change institutional arrangements (particularly if one aim of the new approach is to win worker consent), and traditional sets of expectations and norms may carry over into the new regime. This was found to be the case, for example, in Batstone *et al.*'s study of attempts to introduce comprehensive and systematic reforms (1984).

Traditional structures and strategies may limit the scale, and increase the cost, of change in other ways. The existing structures of management (and unions) are likely to limit the extent to which particular changes are considered and the ease with which they can be carried through. Many features of an organization become taken for granted by participants—indeed this is a condition of co-ordination and day-to-day action. Little consideration may therefore be paid to the possibility of expanding the range of innovation. There may be an assumption of continuity in what are deemed to be non-central areas. Changes invariably affect the various interests within management and unions in different ways; the more powerful are likely to resist accretions of power and influence to traditionally subordinate groups. There will be functional and sectional jealousies (see e.g. Pettigrew 1973; Dalton 1959; Armstrong 1986; Batstone 1979). Given these factors, major changes often appear to be conditional not only upon external contingencies, but also upon a change of key personnel. New leaders seem better able, with outside support, to change the distribution of power and indeed personnel within an organization.

Given these constraints upon action from within the organization, what is often striking is not so much that structure follows from strategy (Chandler 1962) but rather the reverse—that strategy is constrained by existing structures (see e.g. Gospel 1983). There are, therefore, pressures towards seeking change incrementally and introducing modifications to the existing pattern of labour regulation as and when opportunities arise. Moreover, such opportunities will often be less than ideal, so that the initial goals of management are not achieved, and the contradictions and tensions within patterns of labour regulation are increased. This, it has been argued, is a characteristic of many so-called reforms of workplace industrial relations (Batstone 1984).

The preceding discussion has focused upon the forces for stability in patterns of labour regulation and the reasons for the apparent incoherence of patterns of labour regulation. It may therefore be useful to switch from the dominant tendency in much discussion—namely, a static analysis—to a more dynamic approach. The former tends to see rationality and coherence in management action as measured by abstract and largely theoretical models. We have cast doubt upon the adequacy of many of these. A more dynamic approach may indicate the 'situational' rationality of actions on the part of particular managers—that is, given the political and other constraints under which they operate. However, it should be stressed that we are not suggesting that such a political and processual approach would lead to the conclusion that managers achieve their ends. What appears to be a rational strategy on the part of one actor may be countered by another. In addition, unforeseen crises and contingencies may change the effects of actions or, indeed, the priorities of those concerned. This is particularly likely where management is relatively unco-ordinated and places little emphasis upon the question of labour regulation.

In this section we have so far confined the discussion to management strategy. In preceding sections we have also touched upon the question of the extent to which labour issues do in fact impinge upon and shape management's more general strategies. We have noted that labour regulation, contrary to many labour-process theories, is rarely the subject of serious deliberation at the highest levels within management. But frequently the criteria which do figure in debates at this level have important implications for labour regulation. For example, the primary reasons for introducing new technology appear to concern better control of production processes, improved product quality, and cost reduction (Northcott and Rogers 1982). To the extent that levels of output and quality standards are affected by workers, and to the degree that labour is a significant proportion of total costs, there is clearly a link—even if it is not explicitly made—between the general strategy of management and labour regulation.

Managers often hold unarticulated conceptions of the nature and competence of their workers, which affect their approach to labour. This may be particularly true in Britain because management is rarely induced to discuss and debate these assumptions (see e.g. Winkler 1974; Lupton *et al.*

1979; Maitland 1983). It follows that labour issues cannot dominate the general strategies of management. Indeed, labour strategies are likely to be formulated rather late in the day, when options are constrained by a whole series of decisions which have been made already. In this way, therefore, labour regulation issues are likely to be even more strongly shaped by other factors than might otherwise be the case. But it does not follow that the degree of worker regulation is tighter. Indeed, it would seem that the failure to think consciously and coherently about labour matters means that workers are frequently left with a considerable degree of influence, if only because 'working knowledge' assumes especial importance for the successful operation of new systems (Kusterer 1978).

While these tendencies may be particularly marked in Britain, we should expect more generally that policies of labour control derive from the more general strategies of management. By this we mean that the gap between profitability and labour control, which is so strongly emphasized by many labour-process writers, is bridged by the other ways in which employers seek to achieve profitability. Again, this has been stressed by numerous writers. Kelly, for example, has emphasized the complex interlinkages between market pressures, the type of labour recruited, and labour control within the workplace (1985). (Many earlier labour-process writers had sought to evade the question of market pressures by locating their arguments within the era of 'monopoly capitalism', which, it was claimed, permitted companies a high degree of control over product markets.) Especially in view of the focus of this volume upon the question of new technology, we should also wish to stress the way in which the more general nature of the product and the production process affect the sorts of labour regulation strategies which are pursued (see e.g. Jones 1982).

If we see labour control—at least as far as task performance is concerned—as intimately related to the broader nature of the production process, then the work of a number of pre-Braverman writers assumes particular relevance. Notwithstanding some problems in their analyses, the emphasis upon the way in which uncertainty shapes the nature of the production process, found, for example, in the work of Perrow (1970), is of considerable interest. For uncertainty imposes constraints upon the nature of capital equipment and, relatedly, suggests that there will be considerable dependence upon human initiatives and day-to-day decisions. If this is so, then the extent to which it is possible to institutionalize many aspects of labour control is also likely to be limited.

Another older study of relevance in this context is that of Reeves and Woodward (1970). They identified two key dimensions of the systems used to control production. The first was the extent to which control was personal, administrative, or mechanical. The second concerned the extent to which control systems were unitary or fragmented. The interest of the first dimension stems in part from its similarity to the categorization of labour control employed by Edwards. Given this, it suggests that many features which might be seen as labour control should in reality be seen in terms of the control of the production process. Our case-studies highlight a number of examples which support this view. But, once a particular form of production control is introduced, it is likely to affect the forms of labour control. Offe (1976) and others (e.g. SSRC 1968) have stressed the way in which, as production systems become self-monitoring and controlling, traditional forms of close supervision become less applicable, since the key role of the worker cannot be effectively controlled in this way. There is therefore a need to win worker commitment. It does not, of course, follow that this is what management necessarily does. But, if nothing else, such arguments serve to highlight the derivative nature of many aspects of labour control.

The second dimension of the Reeves and Woodward categorization indicates not only that, as noted above, the coherence of management strategy may be variable, but also that the extent to which the broader control system serves to constrain and shape patterns of labour regulation may be variable. As was noted in discussing technology, it has been observed frequently that technology appears to impose a greater constraint upon those features of social organization which are more immediately associated with it. This again may help to explain why work organization often changes with new technology while the pattern of industrial relations does not.

If the pattern of labour regulation is shaped by the nature of other management priorities and, in particular, the nature of the production control system, another reason for the typically limited scale of change associated with new technology is suggested. Unless the new equipment introduces major new principles or is associated with broader changes, there will be little reason for management to seek widespread transformations in the system of labour regulation, including task organization. Indeed, there may be very strong reasons for seeking to minimize the scale of change.

In summary, we have pointed to a number of factors which are likely to limit the degree of coherence in management's approach to labour and have also argued, as have many others, that labour control is related to other policies adopted by management. The extent of such links appears to be variable, not only according to the particular aspect of labour regulation under consideration but also due to the structure and priorities of management.

However, it is also necessary to note that unions may play an important role in shaping features of labour regulation. In our previous volume we pointed to the importance of the scope and sophistication of union organization, along with the level of bargaining and the extent of multi-level bargaining (Batstone and Gourlay 1986). At workplace level, we argued, organizations which covered the labour force as a whole and which had a structure which permitted both the recognition of different interests and their co-ordination were more likely to adopt a strategic approach towards issues and to enjoy a greater range of bargaining. That is, they were more likely to shape the general pattern of labour regulation and to include questions concerning work organization within the system of joint regulation. Such sophisticated organizations were also more likely to engage in bargaining at a variety of levels, particularly where the formal level of bargaining was more centralized. Sophistication was operationalized in terms of the density and institutional security of membership, numbers of shop stewards, and the existence of three forms of co-ordination—senior stewards, full-time stewards, and regular shop steward meetings. This basic approach—which bears strong similarities to our earlier discussion of the organization and co-ordination of management—is also adopted in this volume and needs little amplification at this stage. However, as will be seen, the depth of the case-study data permits us to look more closely at certain aspects of union sophistication, and to investigate more fully the interaction between union sophistication, member commitment, and the general pattern of labour regulation.

In recent writings on the labour process and new technology, particularly those based on case-studies, there has been a widespread recognition of the multiplicity of forces shaping management strategy and the impact of new technology. We have sought in the preceding pages to link the questions raised to more general debates. In doing so, we have tried to point to the need for more systematic analyses, which will facilitate comparison and which may have a wider relevance. Important among these is our approach to union organization and the pattern of labour regulation. In the case-study chapters, however, we will confine ourselves to a largely qualitative discussion in order to highlight the key areas and processes of change. In the final chapter we seek to systematize our findings in line with the discussion of this chapter.

THE RESEARCH

This volume discusses the findings of four case-studies. These were undertaken in 1983 and the first half of 1984. The selection of industries to be investigated was determined by our participation in a larger, five-country project, whose findings have been outlined elsewhere (Levie and Moore 1984). Cases were to be drawn from brewing, small-batch engineering, chemicals, and finance. The actual companies studied were selected through discussions with union officials and managers and searches through trade journals. In all, eight companies were approached. Four of these agreed to co-operate in the research. In another case it was found that, contrary to our initial information, there had in fact been no technical change of any significance. In the sixth case we discovered that another researcher was seeking access to the company, and so we did not pursue our approach any further. In two cases management declined to become involved in the research.

The case-studies employed a variety of methods. Interviews were conducted with a wide range of managers, shop stewards, and workers. These were generally of an unstructured nature, although guided by an *aide-mémoire*. In one case—our finance study—a self-administered questionnaire was issued to about half the union representatives in the company. Extensive use was made of both management and union documents and files—these proved to be extremely important both as a source of data in their own right and as a means of highlighting issues which were then discussed in interviews. In addition we observed the organization of work both in new and old work situations wherever possible (in our insurance study older systems were no longer in operation).

Our work in the chemical company focused upon the effects of the introduction of automatic process-sequence control upon production workers, although we also investigated work organization in conventional dedicated, as compared with multi-purpose, plant. The site on which the research centred was the major manufacturing complex of the company and employed about 8,000 people. In all, thirteen unions were recognized in the production division, representing various groups from labourers to line managers. These were divided into eight negotiating groups which bargained centrally with the company. The largest union was USDAW (Union of Shop, Distributive and Allied Workers), which had achieved recognition in the early 1970s. This union represented, among others, all the chemical and other production workers at the site. These were all members of the same works-based branch.

The brewery case-study concerned the differences in work organization and labour regulation of production workers between an old brewery and a new lager plant, both of which were based on the main site of a large multinational company. In all, the company employed about 8,000 people in Britain, just under half of these working at the site studied. In addition to independent staff associations representing clerical grades and foremen, there were five unions representing manual workers. The largest of these—the TGWU (Transport and General Workers' Union)—was divided into two branches, one mainly representing lorry drivers and the other (the brewery branch), production and related workers. Our research focused on the latter group. Annual negotiations covering all manual workers occurred at company level, although there was considerable bargaining at lower levels.

The introduction of CNC machines in the main spares plant of a large multinational company making sophisticated machines for the food and drink sector was the focus of our engineering study. The plant employed 600 people. Two unions covered supervisors and planners, while the 350 manual workers, most of whom were highly skilled, were all members of the AUEW (Amalgamated Union of Engineering Workers). Annual negotiations occurred primarily at the spares division level, although there was a considerable amount of bargaining at plant and shop-floor levels.

Our fourth and final case-study looked at the introduction of on-line processing of personal lines insurance in a large multinational company and focused upon area offices. The bulk of the 4,000 or so staff were employed in a network of about twenty area offices; about a third worked at company headquarters and in a separate central computing centre. In addition, about one-sixth worked in a specialist subsidiary company. The sole union within the company was BIFU (Banking, Insurance and Finance Union) and the company formed a separate division within the union. Membership of a staff association had been transferred to BIFU in the late 1970s. Bargaining was strongly centralized at company level.

Each of the next four chapters discusses the findings of a case-study. We look at the changes in work organization and the way in which technology changed the tasks which workers were to perform. The factors leading to both change and stability in the pattern of labour regulation and the reasons for the introduction of new technology are considered. In addition we look at the role which the union played in these matters, and go on to show that this generally reflects their more general part in the pattern of labour regulation, which is outlined. Finally, we relate the degree of joint regulation to the pattern of union organization. A final chapter seeks to systematize our findings and relate them to themes developed in this chapter.

2

OCCUPATIONAL HIERARCHIES AND UNION MARGINALITY: THE CASE OF A CHEMICAL SITE

THE company chosen for this case-study was a large multinational with traditional interests in chemical production and processing. In recent years it had developed other related interests. Chemical production and processing were carried out by the company's production division, and took place at a number of sites in the UK. The case-study examined changes in the way chemicals were produced on the main production site, focusing in particular on the effects of automation. At this site, only a small proportion of the manual workers were engaged in chemical production, most being employed in further processing activities and distribution. These various sets of activities were undertaken by separate departments.

In this chapter we look first at the variety of ways in which chemicals were produced on site, highlighting the way in which technological developments had involved building a growing range of functions into the equipment itself. The precise methods employed in the production of different chemicals reflected technological developments, the size of markets for different products, and variations in the complexity of their efficient manufacture. Associated with changes in the technology employed, the occupational hierarchy had gradually become more finely graded. This reflected a trend towards the polarization of workers in terms of autonomy, responsibility, and knowledge requirements.

The second section discusses the limited role of the union in matters relating to the introduction of automated equipment. Not only did it have little to say about the decision to install new plants or their design, it played virtually no role in the detailed aspects of work organization. This was typical of the range of influence exerted by the union, as is indicated in the third section, which looks at the more general pattern of labour regulation. It is shown that the role of the union is largely confined to matters relating to wages and conditions; even where it plays a role in matters relating to work organization, such as recruitment, promotion, transfers, flexibility, and discipline, management still enjoys considerable freedom of action. This section also discusses the significance of the internal labour market for the pattern of labour regulation.

The final section turns to the organization of the union in an attempt to explain its limited role, both in relation to new technology and more generally. It is argued that the relative lack of inter-union co-operation, the lack of systematic co-ordination between stewards within the main production union, and gaps in steward representation weakened the union both centrally and on the shop-floor. The failure to fill all steward positions—a crucial feature of union weakness—is then related to members' perceptions of the role of the union and stewards, and variations in the nature of work and supervisory relations.

ALTERNATIVE PRODUCTION SYSTEMS AND WORK ORGANIZATION

Before looking in detail at patterns of work organization and how they have changed with technological innovations, it is useful to have some idea of the processes involved in chemical production. In principle these are relatively simple. Ingredients have to be combined and treated in various ways to encourage chemical reactions leading to the required product. This typically involves a number of stages. The basic ingredients may in some cases require some form of treatment to render them suitable for entering into the production process. They then have to be introduced in the required quantities at the right time and at the appropriate stage in the production process. The ingredients frequently have to be heated to specified temperatures under controlled pressures; the precision and accuracy of these activities may have a significant impact upon the quantity and quality of the product produced. In addition, the process may require that the mix of ingredients be moved from one stage of the production process to another.

The final product then has to be packaged. Throughout the process, at least with fine chemicals of the kind manufactured at the site studied, detailed records have to be kept of the actions taken at the various stages of production, for both legal reasons and production control purposes. In addition, it is important that the equipment used be kept clean to prevent impurities entering the chemical.

The precise way in which these functions are undertaken varies considerably. In the company studied, for example, four different methods were used. In a few cases techniques dating back to the inter-war period were employed. These techniques were characterized by a very low level of mechanization. Workers added the various ingredients manually and heated them in small boilers to the required temperature; they were subsequently allowed to cool and dry. Workers then crushed the chemical into fine powder, sometimes using small mechanized grinders, and the product was packed. Under this method, the extent to which workers employed mechanical aids was relatively limited; they were in frequent and direct contact with the product, and they totally controlled each stage of the production process.

More mechanized plants were of two kinds—multi-purpose and dedicated. In these plants, the product was transferred by mechanical means from one stage of the production process to another—workers were not normally in direct contact with the product, except when introducing ingredients and packaging the finished product. In addition, there were a variety of monitoring and control devices built into the production system, such as temperature gauges and sight-glasses to check the flow of the product. The plant consisted of vessels, connecting pipes, and valves to control the flow of the product from one stage to another. In some cases, where the production process was particularly complex, further technical aids had been added. In one case, for example, if temperatures rose above pre-set levels alarms automatically sounded and, if corrective action was not undertaken, the process might be automatically stopped.

In these more mechanized plants, then, workers had a greater range of mechanical aids—they did not have to move the product physically from one stage of the production process to another; many tasks were undertaken by mechanical means, although these had to be monitored and controlled by workers using a variety of built-in aids. However, vessels still had to be charged with raw materials, waste products had to be removed, the plant had to be cleaned, and the end-product packaged. Process records had to be compiled from observations made on the plant, and these were also used to assess necessary control measures during processing. A typical control operation, for example, was that involved in controlling the flow of a material into a reactor vessel. A valve had to be opened to allow the flow to proceed, and it was monitored at a sight-glass to judge the rate of flow. At the same time it might be necessary to adjust the temperature in the receiving vessel to ensure that it remained within desired limits.

The major difference between dedicated and multi-purpose plants was that in the former only a single product was produced, while in the latter a range of chemicals was made. Hence, in dedicated plants the links between various vessels and other stages of the production process were permanent; changing them would have been a major task. In multi-purpose plants, on the other hand, it had to be possible to change the linkages between different parts of the production process, and so they were characterized by flexible links between vessels and other types of equipment. Thus one week a process might take place involving five linked reactors, while the following week two of the same reactors might be used to produce one chemical, and the rest another; they would therefore be linked with pipes and valves in a different manner from that used in the first week. These differences between dedicated and multi-purpose plants meant that, first, operators on the latter had to undertake a good deal of rigging and other tasks involved in reorganizing the plant to permit the production of the various chemicals; and second, those on multi-purpose plants required a knowledge of how to produce a variety of chemicals rather than just one.

The fourth type of plant was really a sub-type of dedicated plant. It was characterized by a relatively high level of automation, so that the technology assumed various tasks, which were undertaken by workers on more conventional mechanized plants. In this group of plants, the degree of automation had been gradually increased during the 1970s and early 1980s, so that 95 per cent of the manual control tasks which had previously been required had been eliminated.

In automated plants control of the production process was sited in a central control room and was effected through a combination of pneumatic and electronic remote control devices, whereas in more conventional plants control of the production process had to be undertaken at various places in the plant itself. Initially in the automated plant, control was through a plant diagram or mimic panel in the control room. Subsequently, in the late 1970s, computers had been introduced to perform these control operations automatically; computerization was further extended in the early 1980s.

With the mimic-panel system, an operator could initiate sequences in the production process from the control room by selecting the relevant switches on the mimic panel. This also indicated temperatures and pressures, and the operator had to keep these within the prescribed limits by using the central controls. While certain parameters of the production process were pre-set, operators were still responsible for ensuring that the correct sequence of manual operations (selection and operation of switches) was followed. It was possible to make serious errors, for example by mistakenly opening a valve.

Computerization reduced operators' responsibility for monitoring and removed the possibility of wrongly initiating a sequence of operations. Direct digital control of some process parameters (temperatures, pressures) was introduced, together with automated control over related processes. Thus, if the temperature of a reaction began to go outside the set limits, measures to control this would be automatically initiated, instead of having to be started by the operator. The computer program was also designed to prohibit the entry of illicit commands. Operator control was exercised by using a keyboard and VDU (visual display unit), but it was no longer possible to instruct the system to open a valve that should be closed during the sequence of operations underway.

Other tasks allied to processing remained essentially unchanged. Detailed batch processing records still had to be maintained, and records were initially compiled by hand from the readings produced at the mimic panel. With computerization a batch record sheet was produced automatically, but operators still had to check this and add details from batch record sheets produced by hand to record the processes still done manually.

The chemicals made in the automated group of plants studied were initially made in one of the multi-purpose plants. When production was moved to dedicated automated plants the chemical process was itself modified to facilitate automation. In the conventional plants, raw materials could be added in a variety of physical states, but in the automated plants they were all added in the form of solutions. Thus, new tasks associated with their preparation were created and, as they were not automated, they still had to be carried out by workers, albeit using a variety of mechanical aids and even remote control devices. Furthermore, removal of the final product was not fully automated, leaving tasks connected with drying and bulk packing as manual jobs.

Our focus in this chapter is upon workers directly concerned with the production process. It should, however, be noted that, while with mechanized and automated plants the technology assumed a significant number of tasks which had in the past been done by production workers, the complexity of the equipment served to increase the tasks which had to be undertaken by other groups of workers. Most obviously, with more mechanized systems a greater maintenance workload was involved. With the automated system, the computer programs had sometimes to be modified, and this was undertaken by specialists from outside the production department. In addition, there was a more general growth of chemical specialists as the complexity of products and quality standards increased. However, even if we confine our attention to those directly involved with the production process, while automation clearly involved a reduction in labour requirements per unit of output there was no actual job loss due to the new technology. This was because many other changes took place over the long period investigated. At the time the automated plant was built there was actually an increase in production jobs, since the old plant continued to be used for other products. However, subsequently the total number of chemical operatives fell by nearly a quarter as manufacture of a number of old products was phased out. These changes also meant that, whereas in the early 1970s all chemical operatives had been employed on conventional plants of various kinds, automated plants accounted for about two-fifths of comparable jobs by 1984. There is no overall tendency for automated plants to have fewer operatives—numbers employed depend upon the capacity of the plants and the precise nature of the chemical process.

This outline of the different ways in which technically necessary tasks were undertaken has highlighted how technical methods had progressively assumed many of the tasks which were undertaken by workers on older plants. Mechanization had reduced the extent to which workers were physically in contact with raw materials (and this was associated with a considerable improvement in working conditions), and on dedicated, particularly automated, plants had led to a considerable routinization of many tasks. However, there were a number of counter-tendencies which we discuss below. First, however, it is necessary to consider why management adopted these various methods of production.

The choice of production technologies and working methods

For many years the company had steadily introduced new methods of production, as new techniques became available and new products were developed. Consequently, the vintage of the technologies employed at the time of our research varied widely. Pre-war methods of production continued to be used where the production process was simple and demand limited. But many labour-intensive methods of production were becoming costly, particularly given the lower labour costs of competitors in other countries. Hence production of these chemicals was being phased out, and those that remained were typically used on an infrequent basis, this being sufficient to meet demand. Where demand was somewhat greater, the product was produced in multi-purpose plants, which again meant that production did not take place on a continuous basis. In addition, new products might first be manufactured on multi-purpose plants until the level of demand appeared to justify the construction of a plant dedicated to its production. Dedicated plants were used for those products where demand was sufficient to match the output obtained from continuous production.

This general pattern was modified by other economic criteria relating to the available production facilities. For example, considerations relating to the continuing economic viability of older plants, given a particular level of market demand, might act as a constraint upon the construction of dedicated plants. There was, however, a move in the early 1980s to shift production increasingly to dedicated plants and to close down multi-purpose plants. This move was started when demand for particular products fell. But a resurgence of demand led to a continuation of the operation of the multi-purpose plants.

A central factor in the use of more technologically sophisticated methods concerned the maintenance of high quality and, relatedly, high yields. Some chemical processes, for example, were relatively complex and difficult to control. Hence, as noted above, on some conventional plants more sophisticated monitoring and warning systems had been introduced. However, more central to this concern was the development of automated systems of production. It is useful to consider this decision at somewhat greater length.

The construction of the automated plant basically followed a well-established path. In the late 1960s the company had developed a new product, which was initially made in batches in a multi-purpose plant. However, market trials indicated that a larger market existed than could be satisfied from the multi-purpose plant, thus posing the question of whether to build a dedicated plant. Marketing and financial considerations led to a decision in favour of a new plant and, had automated process controls not been developed at that time, a new conventionally operated dedicated plant would have been built. The decision to install automated controls, made by the production and technical managers responsible for plant design, stemmed not simply from a desire to use the latest technology but also from earlier experience and development work.

The decision to use automated control systems in part reflected general trends in the industry (for a discussion of this see Hirschhorn 1984: 41–7). But it was also guided by the particular experience on the site itself. Production in this sector of the chemical industry had been marked by two trends—an increasing complexity of products and chemical processes, and a partly related increase in the complexity and scale of plants. These two trends, and external factors such as changing raw materials and energy prices, had in turn led to pressure on designers for more 'efficient' manufacturing operations. Technical and production managers had become particularly concerned with the problem of standardizing batch yields, in terms of both quality and quantity. These concerns had led to attempts to modify chemical processes, plant control systems, and work organization.

Attempts to resolve these quality and quantity problems, then, were not confined to a quest for a 'technological fix'. Indeed, across the bulk of the production units organizational changes had been directed to the same ends. First, in some cases additional levels of supervision had been introduced. For example, in one multi-purpose plant, where the production process was particularly complex, shift product supervisors who were specialist chemists had been introduced on all shifts.

Second, supervision and task allocation changed. Until the early 1970s, supervisors had made informal checks on the state of production when shifts changed. Then foremen began to allocate jobs at the beginning of each shift on the basis of a formal shift hand-over log and such production priorities as might be set by management. The compilation of this log also meant that supervisors were more aware of what was taking place on each shift. For it was compiled from batch record sheets handed in by operators at the end of each shift, and from a documented safety check on the state of the plant and processes carried out every two hours by the foreman.

Third, on multi-purpose plants work was further reorganized from about the mid-1970s. Previously operators had undertaken all the stages of work required to produce a particular product. But this was then changed so that they undertook only a discrete stage of the production process and then transferred the materials to another operator, who undertook the next stage. In other words, the division of labour was increased.

Along with this intensification of supervision and a more detailed division of labour, bureaucratic rules were more tightly specified. Up to the early 1960s production tasks were guided by process recipes, but the precise sequence and timing of operations were determined by the supervisors and plant operators. Stricter processing rules had been laid down for dangerous chemicals or processes, but in general production had depended largely on operators' skills and discretion. In an attempt to improve production standards, the processing rules were increasingly formalized into written operating procedures which operators were bound by disciplinary rules to follow. Nevertheless, however precise the operating specifications were made, and however conscientiously operators followed them, decisions such as when to start closing a valve were still made by the operator. He could therefore influence the precise quantity and quality of any batch, not to mention the cost of processing in terms of energy used and waste created.

In order to deal with the consequences of such variations in process control, technical and production managers had already embarked upon a variety of experiments, before the need for the new plant had arisen. In the mid-1960s chemical department managers had experienced great difficulty in producing several chemically similar products in the same plant, and they had decided to experiment with automated process controls to overcome the problem. They devised a sequence control system and built a pilot plant, which demonstrated its feasibility and was therefore installed in the existing plant. The outcome exceeded all expectations: the quality variation problem was resolved; repeatability of the processes between batches was ensured; and the usable output from the plant nearly doubled. This experience clinched arguments about future development in favour of a strategy of seeking to reduce operator executive control over chemical process operations. This strategy had been consciously adopted by the time design of the new dedicated plant began. Thus, the company's design team automated practically all the tasks they could. Nevertheless, a small number of tasks concerned with preparing raw materials, and the packaging and dispatch of the final product, were still to be carried out manually. Automation was not attempted in these areas, partly because of the small scale of these departments. However, a manager involved with the design of this plant from the outset agreed that, had the plant itself been built with warehouses at each end, then material handling tasks could have been automated very easily. Since process operators were required in the raw materials and packaging sections, a small number of tasks on the main plant were not fully automated, partly to provide operators with sufficient tasks to make a job, and partly for safety reasons.

Management, then, had become increasingly convinced that technical rather than organizational means were the most effective way of resolving problems associated with quality and output standards. But, while this had led to the automation of new plants and some partial automation of other plants, where production problems had been particularly acute, it did not lead to a widespread modification of existing plants. In multi-purpose plants it was difficult to develop efficient, highly automated systems, and even in other cases the work required was deemed too costly. In these instances reliance continued to be placed upon tighter operating procedures and close supervision. However, no matter what the precise form of control, the key factors which guided management concerned problems of quality and output standards. These, however, had important implications for the nature of work organization and the content of jobs. In the next section we consider these factors more fully.

Work organization

The preceding sections have shown that, for a number of reasons, management employed different sorts of technology. Central to our purposes is the fact that different types of technology varied in the extent to which they assumed functions which were necessary if chemicals were to be produced. Differences in the degree of mechanization and automation, it was noted, meant that the range of tasks which were left for workers varied considerably, although, at the same time, new tasks might also be required of workers.

While the nature of the technology left particular functions to be performed by workers, it did not in itself determine how those functions were supervised and controlled. We have noted that management had increasingly resorted to closer supervision and tighter bureaucratic rules. These clearly affected the autonomy and discretion of workers. However, the way in which the tasks which had to be done were combined into jobs was also important. Technology, rules, and supervision, while shaping job content, may still leave considerable autonomy, discretion, and responsibility in the hands of workers. This may be something of which management is aware in its planning or may be realized by them only subsequently.

In looking at work organization, then, a useful starting-point is to consider the division of labour on the different types of plant. Above the level of the production workers themselves, there were a number of management grades. Group managers were responsible for groupings of plant, while plant managers were concerned with the production aspects of particular production units. In addition to these line managers, there were a variety of technical specialists whose work related to the production process. In Table 1 we list the various groups below this level.

The first of these is the general foreman, who worked only on days and whose primary tasks related to the organization of the supplies necessary for production. Plant operation and process tasks were carried out by shift foremen and process operators, all of whom worked a continental shift system permitting production twenty-four hours a day, seven days a week. Table 1 shows the structure of occupations and grades, and the ways in which these changed from the late 1960s onwards, as well as variations between different types of plant. Operators on multi-purpose plants had a greater degree of autonomy (even after their confinement to particular stages of the production process), required a wider range of technical knowledge, and operated under less routinized conditions than did their counterparts on dedicated plants (cf. Wedderburn and Crompton 1972). These differences are reflected in the fact that only senior processmen were to be found on multi-purpose plants, most of whom were on the second highest grade for production workers. The majority of operators on dedicated plants were processmen on lower grades (the precise numbers varied between plants). In addition, as noted

TABLE 1. Division of labour and job grades in different types of chemical plants

Type of Plant		
Multi-purpose	Dedicated	Automated
Ungraded:		
General foreman (days)	General foreman (days)	General foreman (days)
		Control room operator (staff post 1974)
Shift foreman	Shift foreman	Shift foreman
Shift product supervisor (early 1980s onwards)		Deputy shift foreman (late 1970s only)
Graded:		
Leading hand (Grade **A**)	Leading hand (Grade **A**)	Control room operator (up to 1974, Grade **A**)
		Assistant control room operator (from early 1970s, Grade **A**)
Senior processman (mostly Grade **B**)	Senior processman (mostly Grade **B**)	Senior processman (mostly Grade **B**)
	Processman (Grades **C–D**)	Processman (Grades **C–D**)
Cleaner (Grade **E**)	Cleaner (Grade **E**)	Cleaner (Grade **E**)

earlier, on one plant, where the production process was particularly problematical, a shift product supervisor also existed.

On dedicated plants no rigging had to be done and, since only one product was produced, tasks which had to be undertaken were of a fairly routine nature—the major variations depended upon the stage reached in the production process. Formerly, when operators started a shift, they had simply carried on with the tasks which their counterpart on the earlier shift had been doing. Later, as noted above, change-overs were more closely controlled by supervision. But while the routine nature of production led to a fairly strict demarcation between the work done by various grades of operator and little variety in work (since in only one case did the foreman permit job rotation), it also meant that processmen were relatively autonomous once they had been assigned a particular set of tasks.

In multi-purpose plants, the greater variety of tasks meant that workers were more likely to be assigned to different tasks not only at the beginning, but also during shifts. In addition, workers might often be assigned to small teams, under the direct control of a leading hand, in order to undertake larger tasks, e.g. those associated with rerigging the plant for the production of a different product. However, operators usually worked alone, since they performed all the required tasks on 'their' section of the plant.

Despite the fact that workers' actions were significantly shaped by the stage of the production process, the operating procedures, and supervision, they had a considerable degree of discretion even on dedicated plants. While breaches of the standard operating procedures could lead to summary dismissal, the rules could be followed rigidly or flexibly according to an individual operator's abilities. Newly trained operators, for example, tended to stay to monitor a process throughout all its stages, and could thus work on only one stage at a time. Skilled operators, however, knew from experience the extent to which a particular part of the process could be left, and could therefore move on to other tasks. Such flexible treatment of the operating procedures led to higher productivity. But it could also lead to errors, and these indicate the degree of responsibility—if only of a somewhat negative kind—which operators had for the final product. In one case, work on a process had to be stopped because of a plant failure and, while repairs were under way, an operator prepared other vessels which would be needed later in the process. The repairs were completed by the time the next shift came on duty, and the operators were able to restart the process. However, they failed to notice the extra preparation work which had been done, and mixed chemicals from different stages of the process. As a result the whole batch had to be repurified. In addition, workers frequently undertook tasks which were not included within their task requirements. These included temporary maintenance work when minor plant problems arose and collecting materials from the stores to alleviate temporary shortages.

In addition to assigning workers to tasks, supervision on the job was limited mainly to checking that safety procedures were followed, that operators attended to the jobs they had been assigned to (and did not spend too much time helping each other out), and that processes were progressing satisfactorily. This meant that process operators were in practice responsible for the specific quality and quantity of product in each batch, as well as for the safe operation of the plant. Product quantity and quality norms were set by management but, as noted in discussing the rationale for automation, small variations in temperature and operating conditions which could be affected by operators could have considerable effects upon the product.

The job structure in the automated plant followed the same hierarchical pattern but with a number of important differences. Of greatest significance was the control room operator, who initially had been graded at the same level as the leading hands on conventional plants. However, it was gradually realized that this job was a particularly important and responsible one. It was accordingly upgraded to staff status and, in addition, an assistant control room operator was appointed (these positions continued to be filled from the ranks of processmen). These more directly task-orientated positions led not only to the disappearance of the leading hand, but also to the elimination of the post of deputy shift foreman (this position was dropped when production was reduced, but it was not reintroduced when output was subsequently increased). Furthermore, this more 'top heavy' structure and the nature of the technology were associated with rather different patterns of working on the automated, as against conventional, plants.

In conventional plants, control and monitoring functions were distributed throughout the plant, and this was an important factor in the relative autonomy and responsibility of the operators. In the automated plants, however, most of these functions were centralized in the control room. The control room operators therefore allocated operators to tasks, and, in addition, through the technical control system, were often able to control their tasks in considerable detail. This new pattern of labour control was symbolized by the use of two-way radios. Through these, control room operators could instruct operators to undertake particular tasks depending on the state of the production process. If the process stopped at a point where a sample was needed, the control room operator would instruct a processman to take the sample. The operator might then perform simple tests on this or set it aside for laboratory testing. There were also a small number of manually operated valves, mainly for waste removal, which were operated by processmen acting on control room operators' instructions. The control room operator not only instructed the processman to undertake this task, but also controlled whether or not the manual valve could be opened. The process operators' tasks were thus continually supervised from the control room: otherwise their work was similar to the simpler work in conventional dedicated plants, except that they were less able to affect the production process. However, they still had some discretion. They were able, for example, to take breaks—as in other types of plant—as and when work permitted, and they would also walk around the plant during the shift to keep an eye on the behaviour of the various items of equipment. Finally, work involved in packaging the product was similar to that found on conventional plants making similar products.

For the typical processman, therefore, one can identify a fairly clear pattern across the different types of technology. As one moves from labour-intensive to multi-purpose to dedicated to automated plants, the degree of responsibility, autonomy, and knowledge required declined. However, it should also be noted that similar trends of reduced autonomy had occurred in all conventional plants, due to the introduction of tighter operating procedures and changes in supervision and task allocation. Against this, however, the control room operator had a great deal more responsibility and required a much greater knowledge than any worker on conventional plants.

The control room workers' tasks centred on the operation of the computer keyboard, which had replaced the mimic panel (these tasks have already been outlined). But the centrality of the control room operator not only meant that he supervised the whole of the production process and assigned workers to particular tasks in the same way as foremen on conventional plants, but that he also assumed other functions which were traditionally those of supervision. The computer programs controlled the timing of processes, but not when they would start. Production had therefore to be planned, and this permitted a degree of flexibility for exchanging tasks between shifts. The control room operator was responsible for these decisions. Towards the end of each shift he prepared a list of jobs which were likely to be necessary on the next shift, and this was used by the incoming foreman to allocate jobs.

Problem-solving was among the most important aspects of the control room operators' work. Like planning, problem-solving first of all required extensive working knowledge of all the equipment and the chemical processes. One operator accurately described the difference between work in an automated plant and that of a skilled operator in a conventional plant as follows: the latter had to know all about one particular piece of plant and its associated processes, while the control room operator had to know all the plant and processes connected with the particular product. This included new 'tacit' skills, in so far as knowledge that a particular valve operated more slowly than it was supposed to (so that the computer registered a malfunction) gave them the confidence to override the computer and proceed with production. In short, in a number of crucial respects the control room operator monitored and controlled the control system (cf. SSRC 1968).

The control room operator therefore became central to the production process. This does not seem to have been fully anticipated by management, since the logic of their resort to technological solutions to the problems of quality and yields was that the role of workers in the production process should be reduced. This was in fact achieved as far as the average operator was concerned. But the importance of the control room operator, who increasingly combined control of the production process with control of related planning and allocative tasks, was only reflected in promotion of the job to staff status some time after the plant had been operating. Ironically, therefore, automation had made production highly dependent upon a key set of operators who had become so important that they assumed many of the functions undertaken by supervision on conventional plants.

In sum, work in the automated plant was organized on lines broadly similar—in terms of the detailed occupational divisions—to those in conventional plants. However, there was a greater division on the one hand between the typical processman, who performed the simplest chemical processing and manual tasks under relatively close control (although some still had considerable responsibility, for example, for highly dangerous chemical processes), and on the other hand the control room operators (and senior processmen who would stand in for them when necessary), who carried out the control tasks. Such a division, however, was becoming more marked on conventional plants, with changes in task allocation, supervision, and standard operating procedures. Our findings therefore suggest that both the optimism of Blauner (1964) and the pessimism of Nichols and Beynon (1977) are one-sided accounts of work experience on chemical plants.

The type of control exercised by control room operators on the automated plant had changed considerably, relative to conventional plants. They no longer controlled detailed aspects or stages of chemical processing—these had been automated—but they now controlled the whole process. Their jobs required extensive working knowledge of all the processes carried out automatically, together with the ability to plan work on a day-to-day basis and to supervise effectively plant process operators. The strategy of reducing operator executive control over process details had clearly been achieved, but it had made one group of operators—those in the control room—even more central to the production process in a way which management had not initially envisaged.

UNION INFLUENCE, TECHNOLOGY, AND WORK ORGANIZATION

The preceding sections have looked at the way in which work organization and content varied between different types of technology and over time, and the reasons for these variations as viewed from the standpoint of management. However, an important question is how far the unions or workers impinged upon these decisions. Unions might influence work organization in a variety of ways: first, they might affect management decisions on the type of technology to be employed; second, they could play a role in shaping the division of labour and gradings of jobs associated with a particular type of technology; third, they might play a role in determining the rules and other controls exercised over workers; and fourth, they might play some role in shaping the way in which controls were exercised on a day-to-day basis.

Many of the existing plants on the site studied had been designed and constructed prior to union recognition in the early 1970s. However, even since then the union had played virtually no role in the choice of technology and associated issues. New plants were typically welcomed by the union as an indication of future job security, and because working conditions in new plants were typically superior to those on older plants. But the union at site level took no initiatives on the choice of technology or work organization, since such matters were deemed to be the responsibility of individual stewards; these in turn rarely played a role, if only because there was not a structure which facilitated such activity. The limits on the stewards' ability to exercise any influence can be illustrated by the degree of consultation which took place over the latest round of innovation in the early 1980s.

Management decided to upgrade the computers in the automated plant, since the original equipment installed in the mid-1970s was out of date and increasingly prone to failure. After the old system had experiencd a serious failure, the union was informed at a departmental committee that management was considering installing new computers. There was no discussion of this plan at the meeting. At a subsequent committee several months later, the union was told that costings for the new computers were being sought; it was subsequently told that these had been received—but not what they were—and that the board had approved the investment. Construction began about two months later, less than a year after the project had first been hinted at.

Although management provided more information in this instance than it had in the past, nothing more than that outlined above was provided, nor did the stewards seek any further details. The general ignorance of stewards is indicated by the fact that they differed in their views as to whether or not they had in fact been told of management plans before construction work began. However, one of the control room operators was asked to set out his ideas for the design of the control room, and saw the plans when they had been drawn up. But, as a member of staff, this worker was no longer a member of the manual workers' union. In addition, one of the process operators' stewards saw the plans at this stage and was able to point out that there was no provision for a toilet in the new unit. In short, there was scant thought given by the stewards to the new equipment, and little use made of the somewhat limited opportunities with which they were presented. To the extent that management consulted workers, they focused upon particular individuals rather than their representatives. More generally, in the words of one manager, 'the technological change was so great that the technical experts dominated the scene'.

Not only did the stewards play a very marginal role in the design of the new equipment, they also played no role in determining the division of labour once a particular type of technology had been introduced. However, operators on the automated plant had attempted to obtain higher job gradings in the early 1970s. This had been stimulated by a manager's statement that above-average operators were required on the plant, but the demand was opposed by the union. Subsequently, however, a senior production manager had suggested that control room operators be upgraded because they often stood in for the foreman so that it made sense to promote them to staff status. This led to their upgrading in 1974. However, even in this instance the union was not actively involved.

This negligible role of the union was reflected in the part it played more generally in relation to work organization—both on the automated and conventional plants. The standard operating procedures, for example, were neither jointly determined nor even the subject of consultation: they were unilaterally written and amended by chemical specialists. Similarly, the union did not discuss manning levels or supervision. Nor were they, as a matter of routine, consulted about changes in working practices. As is implicit in the preceding discussion, the steward's role in day-to-day job allocation was very limited indeed, all allocations being made at the foreman's discretion. Stewards were sometimes involved in dealing with members' queries over job allocation, although this was invariably at the workers' initiative. Typically, stewards raised such problems with the plant manager only if it appeared that workers were being asked to undertake tasks which fell outside the scope of their job descriptions (these were typically fairly broad).

In summary, the role which the union played in relation to the more immediate aspects of work organization, and the strategic considerations which served to shape them, was negligible. However, this is not to say that the unions played no role in such matters. Their primary influence, such as it was, derived from the broader structure of labour regulation, the theme to which we now turn.

THE BROADER STRUCTURE OF LABOUR REGULATION

It has been widely recognized, and further emphasized in Chapter 1, that it is insufficient to look merely at task-related aspects if we are to understand the nature of labour regulation. This is so for a number of reasons. First, the broader structure may consist of a series of rewards and sanctions which induce workers to comply with management goals as embodied within task organization. Second, some aspects of the broader pattern of labour regulation may impinge more directly upon the day-to-day aspects of work— this is particularly true of matters relating to recruitment, promotion, and transfers. Third, the broader structure of labour regulation covers many issues which are of significance for workers' experiences and life chances. Fourth, while many union organizations do play a role in shaping the detailed aspects of work organization, the classic areas of union activity have typically related to the broader aspects of labour regulation. In this section we consider first those aspects of the general pattern of labour regulation which intrude most directly upon work organization, and then move on to consider more general features of wages and conditions. In the chemical company, an important feature of the former concerns a particular kind of internal labour market which generally operated across all departments at the site.

A key feature of management's approach to recruitment to vacant positions, other than those at the bottom of the hierarchy, was internal promotion. A general commitment to this was embodied in the job evaluation agreement, although this had been a policy of management for many years prior to union recognition. Vacancies for operators, for example, were advertised internally and any worker could apply, although by custom leading hands were appointed from among the senior processmen on a plant. Management selected the worker to be appointed to the post, and the unions played no role in this selection process. As will be seen below, the opportunities for moving between jobs were important in that they permitted workers to adopt individualistic strategies directed at improving their job content and pay, and/or to avoid work relationships which they found irksome. New recruits were appointed not only to a particular job but also a particular shift, and they generally remained on that shift for the whole time that they held the job concerned.

This general pattern of internal recruitment was followed when the new automated plant began operations. All jobs were filled by recruitment within the site. The initial batch of operators was selected from among experienced conventional plant operators who had applied for the new jobs. Subsequently, recruits for the control room jobs were selected from among the first batch of plant operators, while recruits to process jobs came from a variety of conventional plants and grades.

However, there were exceptions to this general pattern. In the early 1980s, for example, management had planned to shut down a number of multi-purpose plants, when demand for the products they produced fell. Workers were either transferred to other jobs or, in a few cases, where transfers were unacceptable, made redundant (see below). However, demand then picked up and management, uncertain of how permanent this was, opted to introduce a small number of temporary workers. The stewards became involved in this move—they were unhappy about it but, on being told that no permanent positions would be created, accepted management proposals on the grounds that providing temporary employment was better than providing none at all. (A by-product of this was to introduce further stratification into the division of labour on the multi-purpose plants—the temporary workers were able to undertake only the least skilled tasks, so that lower-grade permanent workers undertook only the more skilled aspects of their conventional workload.)

Generally speaking, the unions played no role in determining either the hierarchy of jobs or manning levels. The degree of stratification of the labour force on the chemical plant, for example, had been progressively increased. This had become a feasible option as production became more routinized and as the technology assumed a growing number of tasks. Job hierarchies reflected the fact that a great deal of the 'skill' of more senior processmen was acquired through experience and was, in large part, specific to the plant on which they worked (this explains, for example, why key jobs such as that of leading hand were traditionally filled from senior processmen). It took several years, we were told, to become fully 'skilled' on a plant. In other words, while the site as a whole was the unit for the internal labour market at less senior positions, that for posts requiring a greater degree of plant-specific skill was confined to those currently working on the plant concerned.

The hierarchy of jobs in the plant also served both to moderate systems of authority and to confuse promotion prospects with the exercise of supervision. Workers often worked alongside those who were formally in charge of them. The legitimation of the latter's position rested less on discipline and direction than upon superior skill, which was exercised primarily in relation to the production process rather than purely in terms of labour control. In addition, the structure of the internal labour market encouraged workers to accept the legitimacy of the structure of authority—the existence of a supervisory position was their route to self-advancement. Hence, despite a system of fairly close supervision in some situations, superior–subordinate relationships were typically relatively easygoing. Such relationships were also facilitated by the small size and relative isolation of many plants.

There were, however, some exceptions to the lack of union influence over these matters. Occasionally, for example, stewards might claim that an additional leading hand was necessary on a plant—but such discussions over manning could scarcely be termed negotiation, nor did they involve any collective pressure. Such proposals also sometimes implied higher manning levels. Issues of the latter kind were not the subject of formal negotiation—management was not prepared 'to become involved in negotiations on manning levels', although this did not mean that 'they would not be prepared to have discussions with the trade union if [management] genuinely thought they had a reasonable point to make about staffing levels'. Union influence was therefore marginal in this area, although senior shop stewards were involved in more formal discussions with management over the redundancies mentioned earlier. In this instance, only one steward questioned management's basic analysis of the situation, but his objections did not lead either to serious negotiation of the situation confronting the plant or to a discussion of specific manning levels. The stewards simply accepted that transfers were preferable to a greater number of redundancies. Negotiations therefore focused on the timetable, principles of selection of individuals to be transferred, and measures to protect transferees against loss of earnings.

Issues relevant to labour allocation also arose in relation to reorganization of work, for example the shifting of products from old to new plants. In one recent instance management had planned that, rather than workers from the old plant 'following' their work to the new plant, the jobs would be filled through the conventional procedures of advertising 'new' jobs internally. This was not the subject of formal negotiation or discussion with the stewards, and the senior steward learnt of management plans only through informal contacts. He raised the matter with management, but could only achieve an agreement that the advertisements for the 'new' jobs would initially be placed solely in the old plant.

Temporary transfers from one job to another, which sometimes lasted several years, were also a common feature of the chemical plants. These were covered by a flexibility and mobility agreement, which specified that if such moves involved a loss of earnings the worker would receive compensation, although at a rate which fell as time passed. The extent to which the union was involved in the selection of workers for transfer varied significantly, depending largely upon the relationship between a steward and management. Many had no influence, but some were able to ensure that transfers were favourable to individuals, in terms of acquiring new skills. In this area, then, the union's role was largely confined to financial compensation; its influence over selection of transferees was patchy (possibly depending upon views about any injustice in management proposals), while its role in identifying the need for transfers was non-existent. This pattern largely reflects its more general role on issues concerning work organization.

A second feature of the broader pattern of labour regulation relevant to work organization was the disciplinary procedure. Under this, management could issue warnings if operators failed to reach 'required' standards of work, and they could punish by suspension, transfer to other work, or dismissal. Under the procedure workers could ask to be defended by a steward, but they were not obliged to be. Stewards were sometimes able to mitigate punishment measures, which were primarily concerned with the way in which, for example, transfers might seriously affect a worker's earnings. Many stewards believed that management had in recent years adopted a harsher attitude to discipline.

A third feature of work organization concerns working conditions and health and safety. Some stewards occasionally became involved in issues of this kind, but the primary responsibility for health and safety lay with plant-level joint committees of workers and management, on which the stewards played no particular role. In addition there were, in accordance with legislative requirements, health and safety representatives, but these were not necessarily shop stewards.

The primary role of the union, then, lay not in the detailed aspects of work organization and related issues, but in matters relating to the rewards which workers received—that is, they focused upon the wage side of the wage–effort equation. Earnings were made up from basic rates, overtime, shift and conditions allowances, and a profit-sharing scheme. The last of these was not the subject of negotiation; workers with a year's service might expect about 5 per cent of their earnings from this scheme, while long-serving employees might obtain up to 15 per cent. Conditions money was a small element and was received by only a limited number of workers—such an allowance would contribute no more than 3 per cent to the wage packet of a low-grade chemical worker. The payment of such allowances had been the subject of local negotiation until the late 1970s. Subsequently it was administered under a more formal arrangement, which served to centralize bargaining. The bulk of workers' earnings thus came from the basic job rate, which accounted for about three-fifths of a low-grade chemical worker's earnings, with shift allowance and overtime each accounting for a further 15 per cent. Controls over these elements were therefore the most important.

Basic earnings could be influenced in two ways: first, through upgrading jobs; and second, through the negotiation of rates of pay for the various grades. Job gradings were formally determined through a job evaluation system which had been revised and systematized in the mid-1970s. With this change, stewards became involved in job evaluation committees which made decisions by consensus. Reassessment claims could be made through stewards, but apart from some anomalous ratings which came to light in the first few years of the current scheme's operation, upgradings were of little significance as a union strategy for influencing earnings.

Basic pay was negotiated at the level of the production division—that is, rates were determined across all the production sites (although in a variety of bargaining units: see below). Up to the mid-1970s manual pay rates in the company were relatively low for the chemical industry. But they were revised significantly with the revision of the job evaluation system in the mid-1970s, bringing the company's rates into the upper half of the industry's wages league. Since that time earnings had roughly kept pace with inflation. A key factor in the company's jump up the pay league, however, was a reference over low pay to the Central Arbitration Committee, initiated by the union but not opposed by the company.

Generally, annual negotiations had been conducted with relative ease and speed, to such an extent that some participants saw the process as largely a ritual. They were typically completed within two days, without much additional informal bargaining or discussion. Once negotiated, agreements were balloted on by the membership. On only one occasion had the membership rejected a pay offer, thus providing the union with an opportunity to reopen negotiations. However, the company made no immediate response and a second ballot was organized, asking members if they wanted to take industrial action. This was rejected and so the initial management offer was accepted.

The union played no role in the allocation of overtime in the chemical plants: continuous working meant that overtime was required only when a worker failed to turn up for his shift. When this occurred, the worker on the previous shift was expected to work a double shift. Overtime rates and shift allowances were dealt with in the annual negotiations.

The union did play a role in a number of other features of worker compensation. For example, in the early 1980s management agreed to a demand by the chemical stewards, which had initially been put forward a decade earlier, that a system of 'pay stabilization' should be introduced. (Until then worker's earnings had varied significantly from week to week according to which shift they worked on.) Management finally agreed to average out these fluctuations. Chemical stewards had also been able to negotiate a clothing allowance on the grounds that clothes exposed to chemicals could not be used outside of the workplace. Management's final offer was more generous than the stewards' claim.

Other ways in which steward negotiating activity affected earnings related to individual problems over such questions as sick pay and discipline. In addition, matters relating to transfers and relief-working had pay implications. For example, when relieving a higher-paid job, workers qualified for higher pay only after they had worked a lead-in period. This applied each time they were appointed to the job. Up to the mid-1970s operators had had to work for three or more days in a week to qualify for relief pay. If, as sometimes happened, they stood in for two days in one week and one in the next, they would not receive any additional payment. Pressure to change this led to a new system, under which operators qualified for higher pay after standing in for three days in any two consecutive weeks. Chemical stewards still wanted relief pay paid immediately, but they had been unable to negotiate further on this issue. A related problem arose in situations where manual workers temporarily worked on staff-graded jobs. Until just before our research, no agreement covered this situation. However, a worker disputed the amount management offered him for working several weeks in a staff job, and this led eventually to an agreement on appropriate rates of pay.

In addition to earnings, the other major area of union influence concerned working hours. In recent years the standard working week had been reduced in two stages from forty to thirty-eight hours, following considerable discussion at local and industry level, and in the face of some management opposition. However, in the chemical department, with continuous shift-working, these reductions required either additional overtime or leave. It was decided that chemical workers on continental shifts should have thirteen 'extra' days' holiday a year relative to other workers.

Shift times in part of the chemical department had been modified by local agreement even before the union had been formally recognized. On some plants, shifts were of variable length in order to fit more easily with workers' convenience.

Recent attempts to standardize the shift system and eradicate these local arrangements had met with resistance to any change without financial compensation. Management did not persist with its move towards standardization.

In summary, then, the broader structure of labour regulation largely mirrored the pattern found in relation to work organization: the unions played little role in matters relating to the allocation and use of labour, its influence being largely confined to issues such as hours of work and pay. Even where agreements existed, e.g. over labour flexibility, management still enjoyed considerable freedom of action. Local bargaining by individual stewards was variable and patchy, depending upon individual workers, stewards, and managers. The 'frontier' of control was limited, insecure, and variable.

Central to the structure of control were not only a detailed set of bureaucratic rules relating primarily to the work process itself, but also the particular nature of the internal labour market. This provided opportunities for vertical mobility, which served to strengthen the legitimacy of authority in the eyes of workers; it suffused and dissipated structures of authority and merged labour control with more production-related tasks; in addition, the opportunities for horizontal (or even, in grade and pay terms, downward) as well as upward movement provided workers with a wide range of opportunities (other than individual or collective approaches to management) to escape from situations which they found irksome. The fact that the union played such a small role in relation to the operation of the internal labour market, as well as in relation to work organization more directly, was a crucial factor in its limited control.

We have also seen that the manning and operation of the automated plant broadly followed the principles found on the rest of the site. No significant attempt had been made to change the broader structure of labour regulation, or indeed industrial relations, as new technologies were introduced. One important reason for this was that under the old system management encountered few problems or difficulties. To understand why this was so we need to turn to the question of union structure and strategy.

UNION ORGANIZATION

In many respects the union representing production workers might appear to correspond with the picture of management-sponsored shop steward organizations which was developed by a number of writers in the late 1970s (e.g. Hyman 1979; Terry 1978; Willman 1980). This is so for three reasons. First, the union became established, in part through management support, during the period of widespread industrial reform. Second, it was in an industry where such sponsorship was claimed to be common. Third, it played very little role in work organization issues. We would certainly agree with the view that the limited role of the union was in significant measure attributable to its form of organization, but we do not believe that this was attributable solely, or even primarily, to a direct and conscious strategy on the part of management. Indeed, the question of recognition led to disputes among competing unions and even strike action. Recognition of USDAW as the sole bargaining agent for production workers at the site resulted from the fact that they were able to demonstrate majority membership and refused to support strike action by TGWU members for similar recognition. More important than management sponsorship as such, therefore, in explaining the limited role of the union, was the detailed structure of what superficially appears to be a well-organized union body. In expanding upon this theme we can usefully employ our concepts of inter- and intra-union sophistication.

Inter-union sophistication

The manual production workers at the site studied were all members of a single site-based branch of USDAW. The union had achieved formal recognition in the early 1970s, after density had grown significantly from about the mid-1960s. In addition, there were a substantial number of other unions representing manual workers in maintenance and other areas, and a number of white-collar unions, including USDAW's staff section. In all, thirteen unions represented workers on the site. The manual workers' branch had no contacts, either formal or informal, with the other unions. In the mid-1970s an attempt had been made to organize a joint union forum on the main site, but this collapsed when USDAW withdrew over disagreements about its function.

One reason for this lack of co-operation was the pattern of bargaining units. Bargaining took place at the level of the production division and therefore covered a number of sites, the largest being that studied. However, the unions were divided into eight bargaining units based on occupational groupings. This structure meant that in formal bargaining production workers did not meet either white-collar or craft unions.

The production and warehouse workers' bargaining unit was the largest, covering nearly 6,000 workers in all. Over half of these worked on the main site, although only a minority of them were chemical workers. The largest group on site, and within the bargaining unit, consisted of those in the warehouses. Production workers from the other sites included in the bargaining unit were also represented by USDAW (each site forming a branch of the union) with one exception. This site had recognized the TGWU before coming under the control of the company. It was only with this union that USDAW had any relationship.

Even in this instance contact was largely confined to formal bargaining occasions. There was no union structure corresponding to the negotiating group, such as a joint union committee. Hence the two unions typically met only annually at the Joint Negotiating Committee (JNC) to negotiate over wages and related conditions, although occasionally other meetings might be held to discuss major changes, such as the introduction of a new job evaluation scheme. In short, at this level union organization functioned almost wholly in relation to formal negotiations.

Initially the union side of the JNC consisted of an USDAW national official, area officials from USDAW and the other union, and the senior shop stewards from each site. In the mid-1970s union organization was altered, and the position of chief shop steward created. Appointment to this position was conditional upon the agreement of all the senior shop stewards and the company; the job effectively involved acting as a union representative across all the sites covered by the bargaining unit. Hence, besides taking part in all negotiations above departmental level, the chief shop steward also liaised between senior stewards both on the main site and other sites.

It is clear, then, that inter-union sophistication was limited, at the level both of the site and of the production division. Such collective co-ordination as there was relied upon formal bargaining occasions and upon the post of chief shop steward, whose responsibilities and coverage were also structured by the bargaining unit. Although, as the only full-time steward and as the only permanent means of co-ordination above department level, the chief shop steward held a relatively influential position, the efficacy of co-ordination was limited by the fact that it was so dependent upon this one individual. If we confine our attention to the site studied, then inter-union co-ordination was non-existent. The resultant divisions within the workforce were one factor which served to limit the range of influence of the unions.

Intra-union sophistication

However, USDAW did represent the large majority of manual workers at the site studied, and might therefore be expected to seek a wide range of influence at this level. One reason for its failure to do so was that, with the development of shop steward organization, the branch became sectionalized, giving each of the major groups within the branch a relatively high degree of autonomy. While facilities for steward meetings were available at sectional level, the extent to which they were actually used depended upon the senior stewards concerned. At the same time, the nature of this sectionalization reduced the role which the branch or any other site-level unit of organization could play in co-ordinating steward activity. These themes will be developed in the subsequent discussion.

In the early 1970s, in the years after formal recognition, the branch meeting had acted as an important means of co-ordinating the various interests within the branch. Relatively high attendances at branch meetings were common, largely because workplace issues were discussed and decided upon. But the role of the branch had declined with the development of shop steward organization. Workplace issues were no longer discussed there, and attendances had fallen to such a degree that it was not unusual to have to call off branch meetings due to a failure to achieve the necessary quorum of ten out of 3,500 members.

Before the union was officially recognized, there had been some union collectors but they had played only a minimal negotiating role. Shop steward organization grew rapidly with recognition, and steward density had steadily increased. Initially, provision had been made for one shop steward to each plant, although in reality a few chemical plants had a separate steward for each shift. By the mid-1970s there were between forty and fifty shop stewards, plus two senior shop stewards.

Shop steward organization was considerably reorganized in the mid-1970s and this led to its superseding the branch. First, the post of chief shop steward was created (and filled by one of the senior stewards). Second, senior steward constituencies were changed from geographical areas to production sections, and their number increased to three. Third, it was agreed that there could be one steward on each shift where the union felt this was necessary. This almost doubled the number of formal shop steward positions. Elections were held every two years for both steward and senior steward positions; there was little competition and such turnover as there was primarily reflected resignations of incumbents for personal reasons or because they were leaving the plant or section.

TABLE 2.   Patterns of shop steward organization

Production group	Steward consts.	No. elected	No. members	Actual steward–member ratio
Chemicals	16	9	300	1 : 33
Secondary processing	32	31	1,400	1 : 44
Warehousing	36	34	1,800	1 : 50
TOTAL	84	74	3,500	1 : 42

The pattern of shop steward organization at the end of 1983 can be seen in Table 2. Variations in steward density were largely a reflection of variations in the size of work units. But what is striking is that a number of shop steward positions remained unfilled. This was an important factor serving to limit the degree of intra-union sophistication. Moreover, the majority of these vacancies were to be found in chemical areas, where almost half of the steward positions remained unfilled.

The sectionalization of branch organization meant that the degree of co-ordination between shop stewards at site level was less than it would otherwise have been. It has already been noted that the branch's role in this respect declined with the development of shop steward organization. But while the post of full-time chief steward was introduced, his responsibilities spread across all the sites within the division, some of which were many miles from the main site. Nor were there frequent meetings of the various USDAW stewards on the site, not even of the senior stewards (although, as is discussed below, there were meetings of stewards at departmental level).

This gap in organization was reflected in the limited amount of bargaining occurring at site-level (and indeed at divisional level—for if bargaining was to become more common at this level, the main site would have had to play a key role). Furthermore, there was little pattern to the bargaining which occurred at site level. Other than the discussion of redundancies and transfers noted above, which cut across departments, discussions at site level were typically confined to attempts to extend institutional gains achieved in one area to the site as a whole. Examples include the payment of overtime to shop stewards for attending steward meetings, and the right to hold meetings of safety representatives in working time. As is indicated in our discussion of the introduction of new computers in an earlier section, such strategic issues were rarely the subject of discussion and negotiation at site level.

According to some shop stewards, one reason for the lack of site-level co-ordination was long-standing sectional differences, particularly between chemical and other departments. Chemical workers claimed to have faced opposition to their proposals on such issues as pay stabilization. They believed that this hostility could, at least in part, be attributed to their relatively high earnings. Certainly, there seem to have been few attempts to develop any overall unity, with the consequence that shop steward organization tended to reflect and reaffirm, rather than overcome, sectional differences.

Hence, while there were no meetings of stewards at site level, such meetings were held at departmental level, although often only quarterly (they could be held monthly). Meetings were chaired by senior stewards and operated on an informal basis. They might report briefly on current events, while other stewards might raise problems concerning their plants. Much of the discussion therefore consisted of an exchange of information; such collective opinion formation and decision-making as occurred was a relatively recent development.

The limited degree of co-ordination at the level both of the site and the department meant that stewards were often relatively isolated. As we have noted, some were able to resolve minor problems concerning pay, but the influence which any steward could hope to have upon pay more generally was limited. This was due to the relatively centralized pattern of bargaining and the absence of collective decision-making on the part of stewards. Moreover, the lack of co-ordination meant that problems relating to work organization were generally treated as one-off and parochial issues, rather than matters which might be of more general concern and significance. As a result, there was little pressure to extend the range of joint regulation centrally to cover these issues. If this had happened, then there would have been a framework for greater steward activity on a day-to-day basis. The possibility of demonstrating the role of trade unions to the membership in their day-to-day experiences might in turn have led to a greater union orientation on their part. As it was, stewards were relatively isolated and had few institutional anchors on which to base their activity—as a result their actions were largely shaped by their members, who demonstrated only a limited commitment to collective means of interest pursuit. The pattern of steward vacancies suggests that this was a particularly acute problem in the chemical plants. It is to this question of member orientation that we now turn.

Member orientation

It was noted above that steward vacancies were concentrated in the chemical plants. In this section we consider why this was so. In doing so we look at member attitudes towards the union and their relationships with shop stewards.

Workers in the chemical plants typically worked in smaller and more discrete units than did other production workers on the site. This explains, for example, the lower potential steward–member ratio in the latter areas. In addition, proportionately fewer workers in other parts of the site were on the higher manual grades. There was not the same finely graded hierarchy and pattern of authority relations as in many of the chemical plants. This, and the more direct relationship between worker effort and output in many areas, meant that superior–subordinate relationships were less easy-going than in the chemical plants; workers also often enjoyed rather less autonomy. Finally, lower gradings along with less shift-working meant that earnings were typically lower.

However, these contrasts between chemical and other production areas can serve only as a partial explanation for the variations in steward occupancy rates. For there were significant variations in these rates between different kinds of chemical plant. All steward positions were filled on multi-purpose plants, vacancies being largely confined to the dedicated (including automated) plants. This pattern had been relatively stable over the years, suggesting that it cannot be explained exclusively in terms of chance or personality factors. It is by looking at members' relationships with stewards and their notions of the proper role of the union that we can understand both the gaps in steward representation and the relatively limited role of stewards at shop-floor level.

The primary role of the union was seen by many members to be the maintenance and improvement of wages, holidays, and other conditions of employment. As we have noted, ordinary stewards had little influence over these matters— indeed, no more than the ordinary union member. Beyond this, some members raised personal questions and grievances with stewards, concerning, for example, being assigned to unpleasant jobs, poor working conditions, pay queries, and discipline. But some also acted independently on these matters, approaching foremen and even personnel officers directly. Some workers rejected steward representation even in disciplinary cases. Moreover, some stewards complained that members often ignored their advice, e.g. against under-taking tasks which were not covered by their job descriptions. (In the event of an accident, stewards feared it would not be possible to defend the worker; however, it should also be noted that job descriptions were phrased in a fairly general manner.)

Most members appeared to see the primary role of the steward as an information channel between management and members. Indeed, it was the desire for information about future changes, rather than simply discovering that changes had happened, that often led work groups to select a steward. In a few cases minor grievances, e.g. over holiday pay, had led to similar decisions. These views on the part of members indicate little commitment to notions of union principles (see e.g. Batstone *et al.* 1977).

Members' concern with product and process changes centred upon job security. On learning that changes were planned, the stewards' first response was to enquire about manning implications, since it was then possible to assess how many transfers, if any, might be forthcoming. Plant and process changes might also have implications for health and safety and general working conditions. But the key feature of this communications role for stewards was that it served to guide individual workers' transfer strategies.

The frequency and significance of changes and the avail-ability of alternative channels of information (and grievance resolution) were the main factors explaining variations in steward occupancy rates and, indeed, the more general role of stewards. First, in dedicated plants, where we have seen that the absence of stewards was most marked, information on changes was less important since work patterns were relatively stable, product and process changes being very infrequent. Second, the high ratios of supervisors to routine operators, particularly in automated plants, facilitated infor-mation acquisition and grievance resolution by individual workers. Such patterns were common even where there were stewards. Indeed, members—both where there were and were not stewards—often expressed the view that they had no need of stewards, and even no time for the union, since they could resolve problems by approaching management directly. In some cases, where managers were very approachable and sympathetic, members had decided that they had no need of a steward and so ceased to select one.

However, while members were relatively individualistic, they had on occasion imposed collective pressures on the company. For example, in a dispute over conditions money, a boycott of part of a plant was imposed and nearly brought production to a halt. In another case, all the chemical workers on a shift had applied for casual leave on the same day in an attempt to stop management removing a pay anomaly which worked in their favour.

The role of stewards in such rare events varied. In the first example cited, they appear to have played no role. In the second case, stewards had stressed to members the need to impose pressures upon management. A degree of steward leadership was therefore possible, at least on occasion. However, the second example also suggests the importance of steward co-ordination for such leadership. This was necessary to ensure common action across all plants; such steward unity and mutual support may also have been important in giving individual stewards the confidence to take such initiatives in relation to their members. We have seen that these two conditions of steward leadership were not normally present.

As in all workplaces, and even where shop steward organization is strong, workers were prepared to undertake collective action only on certain issues. The number of occurrences of collective action at the site was too small to make generalizations with any confidence. Simple instrumentalism does not appear, however, to be a sufficient explanation, if only because one would then have expected collective action over a number of other issues. What may also be important is some notion of 'bad faith' on the part of management, e.g. by seeking to withdraw from an agreement freely entered into, or by appearing to act unfairly or in an excessively penny-pinching manner. The rarity of strike action therefore would appear to be due to a combination of two elements—the relative fairness of management action, and the fact that members lacked 'bargaining awareness' and a greater consciousness of the links between individual concerns and collective interests. The latter, at least, can in part be attributed to the weaknesses of steward co-ordination and support.

In this chapter we have seen that the tasks required to produce chemicals could be undertaken in a variety of ways. Managements had gradually resorted to increasingly mechanized and then automated methods of production. These changes meant that tasks formerly undertaken by workers were assumed by the equipment itself, including, with automation, many monitoring and control functions. This meant, first, that the autonomy, responsibility, and required knowledge of many workers was reduced; this was reflected in a greater number of lower-graded jobs on dedicated and automated plant. Second, the requirements of other jobs became greater—this was most strikingly the case of those working in the centralized control rooms on automated plants. They not only needed the knowledge and ability to control the whole plant, but also assumed many planning and labour allocation functions which were undertaken by foremen in conventional plants. The role of the foreman therefore became less central. Third, and related to these developments and other attempts to exert greater control over working practices, the job hierarchy became elongated.

In the main, the changes associated with automation reflected the logic of existing management practices concerning work organization. Equally, there were no attempts on the part of management to change the broader pattern of labour regulation as new technology was introduced. The union played a negligible role in the introduction of the automated plants and this reflected the more general pattern of union activity. Its influence was largely confined to monetary aspects of the wage–effort bargain. It played little part in matters concerning work organization and the internal labour market. The limited role of the union was related to its organization. Inter-union co-operation, at least as far as USDAW was concerned, was non-existent at site-level. Intra-union organization also suffered from a lack of central co-ordination and gaps in steward representation in dedicated chemical plants. They were largely attributable to the dominant definition of the role of stewards as sources of information which could guide individual transfer strategies and as handlers of minor grievances. In dedicated plants the former was of limited significance since work patterns were stable, while grievances could more easily be handled personally, given the structure of authority relationships. More generally, the definition of the role of both stewards and the union indicates a limited degree of 'unionateness' among members; this contrasts markedly with theories of the new working class (e.g. Mallet 1975). Instead, our findings support the basic thrust of Gallie's argument (1978) concerning the importance of what we have termed the broader structure of labour regulation.

In summary, then, one can identify three interrelated factors which served to limit the role which the union played, both in relation to new technology and more generally. The first of these was the structure of labour regulation and work organization; not only was the company a fair and traditionally paternalistic employer, but also it provided important avenues by which workers could pursue indivi-dualistic strategies; second, there were gaps and weaknesses in the organization of stewards; and third, the members demonstrated little commitment to the union or to what have been termed union principles. These three elements may be seen as mutually interacting. However, this is not to say that where such internal labour markets exist there is inevitably little part which the union can play. This and the quite different effects of a similar process of automation form important themes in our discussion of the brewing case-study.

3

SINGLE-GRADE WORKING AND UNION SOPHISTICATION: THE CASE OF A BREWERY

I N this chapter we look at the introduction of a new lager plant at the main brewing site of a multinational company with extensive interests in the food, drink, and leisure industries, and in retailing. Despite this range of interests, brewing was still the company's main source of income. Competition in the brewing industry had forced diversification of interests both inside and outside the industry. The decision to build a lager brewery, the focus of this case-study, was part of this diversification strategy, in so far as it involved a new product which managers felt would enable the company to move into the expanding sectors of its traditional markets.

This study has certain similarities with the chemical case discussed in the preceding chapter. Chemicals and brewing are generally both classed as process industries, and in both cases decisions about new products preceded the introduction of new production equipment. However, the similarities end there—or, rather, these similarities provide the basis for some interesting comparisons. In the brewery, technical innovation was initially conceived with little concern over the labour process and work organization, while the priorities in the design of the chemical plant included improving yields and this had very direct implications for operator control. Furthermore, the brewery union was historically well established, and organizationally very sophisticated, a marked contrast to the union in the chemical company.

In the first section of the chapter we look at the reasoning underlying the construction of the new lager plant and its design and go on to compare the new and old breweries in technical terms. We then compare the different manning arrangements and working practices on the old and new plants and look at how it was that single-grade working came to be established in the lager plant. In doing so we also look at the role which the union played in this matter. This is found to be significant relative to the chemical case, and the next section goes on to look at the more general role which the union played in the pattern of labour regulation. After considering the history of productivity bargaining at the site, the final section shows that union organization on the site was sophisticated and that it was this which permitted it to exercise the degree of control it did.

NEW TECHNOLOGY AND WORK ORGANIZATION

At the site studied, traditional ales had long been produced using only one brewery which had been constructed over fifty years ago. Only a small number of modifications had been made to the plant since its initial installation. However, sales were falling in the early 1970s. This led to two moves on the part of the company. The first was to diversify its activities away from brewing. The second was a decision in the mid-1970s to enter the lager market, which, unlike its traditional markets, was growing.

Market forces were therefore the key factor in management's decision to construct a lager plant at the site. The actual design and construction of the plant were guided by two key considerations. First, since there was some uncertainty about both the extent to which the lager market would grow and what share of that market the company would gain, it was decided to construct a lager plant which would initially have a capacity only one-eighth of the old brewery's, but with scope for subsequent expansion if the market appeared to warrant it. Second, senior management believed that it was vital to launch its lager as soon as possible in order to establish itself in the market before other companies did—for several others were known to be planning similar moves. It was therefore deemed necessary to construct the plant quickly and to minimize the chances of any technical problems which might delay the commissioning of the new plant. Accordingly, the board decided to avoid the application of the latest technical developments in brewing, which had not been fully tried and tested, and opted instead to employ methods which had been thoroughly proven. In the context of the mid-1970s this meant acceptance of 'hard-wired' electro-magnetic switching systems (operated from central control panels) and automated in-plant cleaning. Subsequently, however, computerized process controls were incorporated into the plant design.

The technology to be employed in the lager plant therefore reflected general developments in the industry rather than any technological innovations developed within the company itself. Indeed, the company was dependent on outside expertise, so that the design and construction team set up consisted both of a lager project group from within the company and a number of specialist contractors. It was one of the latter who initially proposed automating process controls, several months after the team had been established, at a time when construction work had already begun and a good deal of equipment ordered.

The contractor appointed to design and install the in-place cleaning system and pipework planned to employ a computer to control the cleaning system, as had become the norm at that time, and pointed out that there would be spare computing capacity which could be used for process control. Some of the company's own managers on the project team had previously favoured making some use of computerized controls, and seized this opportunity to gain experience of these for the company. The extent to which automation could be applied, however, was limited by the financial constraints on the project imposed by the board. Accordingly, it was decided to automate a limited number of valves, mainly those used frequently or which were particularly inaccessible, and also to computerize the plant monitoring system. These plans went ahead, but—despite the stress upon an early start to lager production and the consequent decisions over design—a series of technical problems in commissioning the plant meant that it was not officially opened until 1980, nearly five years after the decision to construct the lager brewery had been made.

The production process

Before going on to consider the division of labour on the old and new breweries, it is useful briefly to outline the basic production processes. These are broadly similar for both the traditional ale produced and lager, although they differ in detail. The production of beers in general involves mixing hot water and malted barley, filtering out the grain husks, and then adding hops and yeast to the resulting liquid. After fermentation the yeast is removed and the beer is (nearly) ready for consumption. The process of adding hot water to the barley can be carried out in two ways, by infusion or decoction. Infusion simply involves adding hot water to the grain, and allowing it to steep. Decoction involves heating the grain and water mixture up through several stages, during which the temperature is controlled precisely. Infusion processing is adequate for most beers, but decoction is essential for lager production. The other main process contrast concerns the end of the brewing process. While all beers are filtered and treated to ensure their quality, ordinary beers are kept for only a few days before being distributed for consumption. Lagers, on the other hand, are stored at specific low temperatures for three weeks or more while secondary fermentation takes place. Lager production thus requires stricter control over temperatures both during and after the brewing process.

The company's old brewery could not be modified to meet lager processing requirements without extensive rebuilding, and so new production equipment had to be installed. However, this was kept to a minimum, and the company built only a new brewhouse and a lager maturation block. The lager plant made use of all the other facilities of the old brewery—milling and raw materials, by-products disposal, production services, and distribution. This did not alter tasks required in those areas, and new forms of work organization were confined to the new brewery and maturation block. The following discussion therefore focuses upon production workers in the comparable departments in the old brewery— brewing and the vathouse—in order to compare modes of control over work organization and union influence. We turn now to an outline of the production tasks carried out in the old and new plants.

In the old brewery virtually all production tasks were carried out by direct manual operation. The addition of prepared grain to hot water at the beginning of the brewing process was effected by starting and stopping a conveyor belt from the grain store. The water was heated by manipulating steam valves—the method used throughout the old brewery, where vessels had to be heated. Monitoring this stage of the process involved observation to ensure that the water and grain mixed adequately, and that the colour of the resulting mash was within the normal range. When this had been achieved, the grain husks were separated from the mash by filtration. This was a process requiring constant monitoring and intervention to ensure that the filters did not get blocked, and the maximum amount of mash was obtained. The mash was then transferred to other vessels by manipulating valves manually and the hops were added. Here it was necessary to measure the volume of mash in the vessels, and this was done using a dip-stick. The hops were added by hand through a charge-hole at the top of the vessel, in quantities varying with the amount and colour of the mash and the type of hops. The hop soaking process also had to be monitored to ensure that it proceeded correctly, and this was done by observation through the charge-hole. When it was judged that this stage of the process was complete, the hops were removed and the liquid passed on to the next stage, addition of yeast. Before this was done, the sugar content of the liquid had to be measured to provide the statutory specific gravity declarations for Customs and Excise and to determine in what proportion the liquid from different mashing processes should be blended to produce the desired quality of product. After fermentation, which was allowed to proceed without any temperature controls, the yeast was removed by centrifuging the beer. The yeast was then pressed to remove any remaining beer and stored for reuse in the brewery, or sent to the by-products department. The beer itself was pumped into storage tanks.

The main process carried out in the vathouse was the 'make-up'—the blending of beer from different brews and the addition of other ingredients to produce the final product. The quality and characteristics of beers from different storage tanks were tested and calculations made to determine the proportions in which different brews should be blended to meet established quality standards. Beers from different storage tanks were then pumped into the vathouse racking vessels in the relevant quantities. The volumes of liquid in the racking vessels had to be measured with a dip-stick through the top of the vessels. Some special beers had to be kept at particular temperatures for up to a week before they could be dispatched, and these temperatures had to be monitored. Other beers had to be pasteurized before final dispatch, and this was also done in the vathouse. This process was virtually automatic since it only required the pumping of beer through the pasteurizing equipment. After 'make-up' and pasteurization, the beers were pumped to the dispatch department, during which process the final checks on the clarity of the product were made by observation at a sight-glass.

Since brewing is a biochemical process, cleaning equipment to prevent biochemical infections is a very important task. All the equipment in both the brewhouse and the vathouse was cleaned each time it had been used. For some vessels this simply meant they were flushed out with clean water, but others required regular cleaning with steam and occasional cleaning with detergents. In some cases equipment had to be partially dismantled (to permit workers access) and manually scrubbed down. Some routine cleaning had recently been automated to the extent that clean water or steam flushing could be effected from a control panel next to the equipment.

In the new lager brewery, mechanisms and methods similar to those in the old brewery could be seen alongside fully automated ones. The first stages of production—the mixing of water, grain, and hops—were controlled by the operation of buttons on the control panel. Nearly all the tasks necessary to carry out this stage of production could be undertaken at this panel, which also displayed the volumes of liquid in the vessels. However, the quality of the final product being transferred for brewing was still checked visually through a sight-glass in the pipeline, while the hops were also added by hand. In this part of the plant, the brewhouse, cleaning was still done in the conventional manner. In the second section of the lager plant, the maturation section (equivalent to the vathouse in the old plant), the remainder of the production process—from fermentation through filtration to transfer to storage, and thence out of the brewery—was completely controlled by a computer in a separate control room. Here the state of the plant as a whole was displayed on a mimic panel with lights to indicate, for example, material flows and whether particular pumps were operating. Other instruments on the mimic panel showed the temperatures and pressures of different vessels, which were monitored and constantly checked by the computer system. Transfers of materials between vessels, however, were not effected simply by means of the computer program. An operator had to set up the pipeline route from one vessel to another by opening and closing the relevant valves along the chosen route, the validity of the route being verified from indications on the mimic panel. When this had been done, the transfer could be initiated by keying in the appropriate code at a terminal in the control room. This operation opened valves to release materials from a vessel, or started a pump to force the material along the route, and these items of equipment were automatically shut down when the operation was complete. Vessel and pipeline cleaning in this part of the lager plant was wholly automated; it required only the keying in of cleaning sequence codes for particular parts of the plant for this to take place. On the other hand, filtration to remove excess yeast from the brewed lager was largely a manual process, the setting up of filters and the control of the flow through them being done by hand, as in the old brewery.

The division of labour

The actual brewing process involved only a small proportion of manual workers employed at the site. Those sections of the old brewery undertaking functions comparable with the lager plant employed about 100 workers, while the lager plant had only a tenth of this number. Given the differences in capacity and the fact that the old brewhouse worked a continuous four-shift system, whereas the whole of the lager plant was only on a two-shift system, manning per unit of output was in fact higher on the new plant than the old one. However, strict comparisons are complicated not only by the factors just noted, but also by the fact that the lager process was more complex while, on the other hand, the plant was more compact. Given the automation of many functions, it seems reasonable to suggest that manning on the lager plant was less than if more traditional methods of production had been employed. But since the plant was an addition to the site, it led to a small increase in total employment compared with what manning would otherwise have been.

The division of labour differed dramatically between the old and new breweries. Fig. 1 shows the occupational divisions in

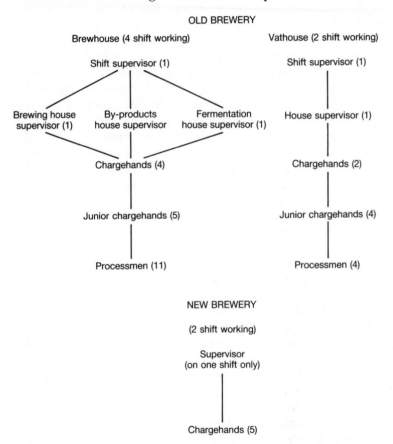

Note: Manning levels varied significantly between shifts, particularly in the old brewhouse.

Figure 1. Occupational structure in comparable areas of the old and new breweries

the two plants. While there was a relatively complex occupational division in the old brewery, both vertically and horizontally, the structure was much simpler in the lager plant. In the latter, all workers might work in any section of the plant, whereas in the old brewery there was a clear division between the brewhouse and the vathouse, and in the former there was a further differentiation among supervisors. Second and most importantly, there was only one grade of worker—chargehand—below the most senior supervisor in the lager plant. Third and consequently, workers on the lager plant had no supervision for half their working time. These contrasts suggest important differences in working methods, and it is to these that we now turn.

In the old brewery, there was a division between the brewhouse and the vathouse. In the former, where workers worked either in the brewing area, the fermentation block, or in by-products, the shift supervisor was responsible for assigning them to their tasks on the different sections at the beginning of each shift. He also planned manning arrangements to cover for holidays and production peaks and assigned workers to relief duties. In addition, the shift supervisor received information and instructions from higher management concerning production plans and was responsible for ensuring that they were carried out. House supervisors were responsible for assigning individuals to particular tasks on each shift, and they also compiled a handover book at the end of every shift for the incoming supervisor to work from. In addition, in the brewhouse they took samples and performed simple chemical checks to monitor the process, and also made out the statutory Customs and Excise declarations. The most important aspect of their job, however, was blending the product of different brews prior to fermentation.

The three process operator grades corresponded to jobs of differing responsibility and skill. The chargehands performed three key jobs—the initiation of the brewing process, removal and processing of yeast, and sterilization of the fermentation tanks and pipelines. They rotated between these jobs roughly once a week, on the supervisor's instructions. The junior chargehands assisted the chargehands in these tasks, but were also responsible for maintaining the water supply to the brewery. Ordinary processmen did the remaining, mostly manual, jobs, and had little to do with process control; they were also rotated around the different duties which were required.

In practice, there was a good deal of additional flexibility in working arrangements. Lower-grade operators frequently stood-in for higher-grade workers and had typically been trained for these tasks. Generally, there were more supervisory tasks than could be performed by house or shift supervisors, so that a chargehand was 'made up' and this led to a sequence of further 'upgradings'. There was thus a considerable amount of vertical as well as horizontal flexibility in the normal routine of work.

The hierarchy of grades in the vathouse was the same as in the brewhouse, although only two shifts were worked here. The shift supervisor performed the same labour planning functions as his counterpart in the brewhouse, and was also responsible for planning and monitoring all beer movements. These were outlined by the vathouse manager, but the supervisor had to plan the movements in terms of the vessels and labour available. The house supervisor was primarily concerned with the technical aspects of work: he was responsible for taking samples, blending beer from different brews according to detailed instructions, and for compiling full blending and transfer records. The need for care in blending also meant that he exercised closer supervision over some aspects of the operators' work: for example, he would check beer transfer routes before sanctioning their use.

The vathouse chargehands and junior chargehands worked in small teams. The chargehand was responsible for carrying out the blending operations and monitoring the beers being transferred to the tanker station. He maintained records of the temperature, clarity, and volume of beer transferred at half-hourly intervals throughout the day. The junior chargehands assisted in these tasks, as well as undertaking other tasks according to the chargehands' instructions. These two grades worked in a complex shift rota, since the types of work performed generally varied between the morning and afternoon shifts. They also swapped shift duties informally, telling the shift supervisor of their intentions. The vathouse processmen worked directly under the instructions of the shift supervisor, and were mainly involved in cleaning vats in preparation for new transfers of beers. A similar pattern of reliefs for higher-graded jobs existed as in the brewhouse, although few of the processmen had been trained to do the work of the junior chargehands.

The organization of work in the lager brewery differed markedly from that in the old brewery. Production—by batches—took place for only sixteen hours a day, and so there were only two shifts. All process operators were graded as chargehands and they all performed the full range of production tasks—from operation of the control panels to sweeping up around the plant. In the control rooms the chargehands controlled the first stages of lager production, monitored the progress of brews, organized routes for trans-ferring materials between vessels, and planned and initiated cleaning routines in the maturation section. The last of these tasks was particularly important, for cleaning had to be done in an order which did not obstruct future planned stages in the brewing process. The chargehands also set up and ran filters and generally ensured that items of plant were operating satisfactorily.

The allocation of tasks between chargehands was made entirely by informal agreement among themselves. An operator who had been in the control room for some time, for example, might hand over to someone who had been setting up transfer routes, and the first chargehand might then operate the filters, sweep up, or make tea. The supervisor played no role in the allocation of tasks, and only occasionally intervened to correct or modify the performance of particular tasks. The only differentiation between operators as far as task performance was concerned was that there were occasional difficulties, for example, when newer members of the team worked in the control room. (When this happened, the work group ensured that a more experienced worker was close at hand to help with any problems which might arise.)

There was only one supervisor in the lager plant, who varied his hours so that he could cover more than one shift each day. His main tasks centred on record-keeping and dissemination of processing instructions to the chargehands, keeping the stores, and generally monitoring production. If a crisis arose while he was on the plant, then he normally took control. He usually arranged for relief transfers of operators between shifts when the workload required extra hands. He also instructed operators to transfer batches of lager to the maturation tanks, to prepare them for dispatching, and to add special ingredients to brews. Otherwise, he had little to do with the daily routine of production. Unlike the chargehands, the supervisor had been trained to program the computer, but it was rarely necessary to make any alterations and, if major computing problems arose, outside contractors were called in.

The nature of the technology and the differences in work organization meant that the operators in the lager plant required greater operating knowledge than even the highest-graded worker on the old brewery. They had to be familiar with the function and position of all pipework, valves, and flowplates over a much longer process than any operator (or supervisor) in the old brewery, since sometimes the plant had to be operated without the aid of the computer (for example, in the event of its breaking down). Second, not only did they have to be able to work effectively at the control panel, but this also required that they understood the process conceptually. Third, the size of tasks given to the operators was typically larger than in the old brewery, since they were generally given, say, a full day's cleaning programme rather than more specific instructions. These greater demands were also indicated by the fact that six months' training was given to operators, considerably more than to any grade in the old brewery.

The mental demands of work in the lager brewery were also greater. In addition to having to operate the control panel, they also undertook a number of tasks which in the old brewery were undertaken by supervisors. For example, they completed the Customs declarations. This involved relatively complex calculations and entries could not be altered; errors could be costly to the company or lead to prosecution. In short, on most counts the requirements made of workers on the lager plant were greater than those made of workers on the old brewery. Their work, however, involved less physical effort and no night-working was required. Nevertheless, it should also be noted that the lager chargehands had to undertake more mundane tasks than chargehands in the old brewery, e.g. sweeping up.

In sum, there were major differences in the organization of work in the new and old plants. In the latter there was a fairly strict occupational hierarchy, although there was some vertical and horizontal mobility. In contrast, the lager plant approximated to a semi-autonomous work group, with all operators being graded as chargehands, arranging task allocation among themselves and often working without any supervision. Hence, whereas in the chemical study it was found that new technology was associated with an elongation of the occupational hierarchy and tighter control over many operators, the reverse happened in the brewery. The obvious question is how this dramatic change in working practices came about.

Decisions on work organization

Despite the fact that, at the time of the construction of the lager plant, there was a great deal of discussion in the labour relations world of the presumed virtues of semi-autonomous work groups and other 'new forms' of work organization, these played no identifiable role in management thinking. Indeed, to a significant degree, the pattern of single-grade working came about by chance. In planning the lager project, management's concern over labour-related issues did not focus upon working arrangements in the new plant. Rather, they centred upon two other issues.

When the company first publicized their intentions to construct the lager plant, they linked formal sanctioning of the project by the board of directors, and thus finance for the project, to union agreement on two points. First, they insisted that the unions agree to negotiate over a package of changes in working practices in the old brewery aimed at raising productivity. This reflected the aim of the company not only to move into new areas of beer production but also to become more cost-competitive in its traditional market sector. Second, the company sought an assurance that, in the event of a strike by lager plant construction workers, the site unions would ignore any picket lines—thus ensuring that work in the old brewery would not be disrupted.

The joint union committee eventually agreed that they would not automatically observe any picket line set up by the construction workers. Their approach on this question would be determined by their unions' instructions. In other words, they did not give management a blanket commitment on this issue. However, agreement to negotiating a productivity deal took a long time to reach. The TGWU, representing the production workers and the bulk of manual workers on the site, welcomed the plans since they felt there was no alternative if the future of the brewery was to be secured. Not only did they feel that economies had to be made if the traditional ales were to remain competitive, but they also saw the lager plant as a symbol of continued company commitment to production at the site. Other groups, particularly the maintenance tradesmen, were less willing to discuss changes in working practices. However, after a good deal of debate within the unions, they eventually gave way. Since it was the practice that issues such as this had to be agreed by all the manual unions, the board delayed issuing its final sanction to the lager plant for several weeks. Nevertheless, the project team continued developing its plans during this period and the unions' actions had no influence on its work. But disagreements over the details of the productivity deal led management to refuse to engage in any formal discussions over the labour aspects of the lager plant for some months. However, informal 'off-the-record chats' did take place and played some role in shaping management thinking.

Management did not seriously consider the manning aspects of the lager plant until its construction had already been under way for several months. The initial proposal necessarily included something on this question in order to permit cost projections; but the figures used were purely nominal, reflecting traditional manning practices, and no consideration was given to how the labour force was to be deployed. However, the designers of the equipment stressed in their promotional material the way in which new technology subsumed control functions previously undertaken by workers (see e.g. Davies 1986: 131). But, as will be seen, this did not influence management thinking. Indeed, by rejecting the latest technology they effectively rejected this as a primary goal.

Some time after the project had been approved, the leading member of the company's lager project team put forward two sets of manning alternatives, differing only in detail, which represented attempts to replicate the form of work organization which existed in the old brewery. First, it was proposed that the brewhouse and the maturation section should be worked as separate departments. Second, there were to be three grades of operator. This number of grades was seen as necessary to permit a separation of jobs seen as 'skilled', to be undertaken by a chargehand, from more menial jobs, which were to be carried out by processmen. In addition, there would be two supervisors, one of whom would operate the control panel. The manning arrangements then became further complicated since, as under the old system, manning levels had to provide for reliefs of the appropriate grades.

These plans were amended following discussions with the personnel department, which, in turn, had discussed the question informally with the union. The key change emanating from these discussions was that the full three-level hierarchy no longer had to be applied. Hence it was proposed that the number of ordinary processmen be increased and junior chargehand posts be entirely removed. However, there still existed problems in aligning different grades with the workloads which were likely to be required.

The catalyst to a resolution of this problem was the decision to introduce a centrifuge (to assist the separation of yeast from the brewed lager), which was made about three months later. This altered the tasks which had to be undertaken by the workers on the plant, and the lager team managers decided it required a chargehand to oversee its operation. They further decided that, since this would increase the number of chargehands required, the division between the brewhouse and the maturation section could be eradicated as far as working arrangements were concerned, and that a charge-hand could operate both the centrifuge and the control panel. At this stage it was planned that there would be seven chargehands, who would move between working in the brewhouse, operating the control panel, and operating the centrifuge. In addition, on one shift there would be four ordinary processmen and a supervisor, responsible for the whole brewery and working only one shift. Outline job descriptions were drawn up over the next month on the basis of this plan.

The TGWU first began to consider its attitude to manning in the lager brewery at the time that the first management document was drawn up and, as noted above, made some informal contribution to management thinking. However, the union was very anxious to open formal talks on manning and training, but management refused to do so until agreement was reached on the productivity deal. As a result, the management team completed their plans without any formal contact with the TGWU. More formal discussions did occur, however, between the TGWU branch chairman and the senior member of the lager project team a month or so after management's plans had been completed—even though no agreement had been reached on the productivity deal. These meetings largely took the form of an exchange of views—no decisions emanating from them were reported to the branch committee, nor did the latter ever formally ratify the plans.

The branch committee was in favour of rotating jobs round a single grade of operators, but was less happy with the idea of having an administrative supervisor. Leading stewards had long been convinced that small groups of processmen could easily supervise themselves and work directly under a manager. This view had previously been put to management informally. However, management was equally under pressure from the supervisors' association to create a supervisory position, as well as being already committed to the idea themselves. The TGWU lost this argument and a supervisor was appointed to the lager plant a few months later. The union, on the other hand, welcomed the lager team's proposals on job rotation among chargehands—an issue where their views had played a role.

As noted above, the lager project team planned to have seven chargehands and four ordinary process operators in the lager brewery when it was fully operational. Initially, however, only five chargehands were to be recruited as the plant was to be run on a single shift for the first few months. As production increased, so two-shift working would be introduced and the full complement of operators recruited. The first five chargehands were recruited some three or four months after the supervisor and began their training. Subsequently two more chargehands were appointed, and the plant was operated by seven chargehands—and no processmen—until the early 1980s, when three more charge-hands were taken on. By the time of our research this pattern of manning still remained. In other words, there were more chargehands than management had previously intended but there were no processmen: this part of the plan had been quietly dropped. Overall, manning was one person less than had been previously proposed.

The lower level of manning is attributable to two main factors. First, the growth of the lager market was not as great as had been expected. This led both to the indefinite postponement of the plans for further expansion of the lager plant and also to levels of working which were well below the full capacity of the plant. Hence, less labour was required. Second, this was acceptable to the unions for a number of reasons. They accepted the need for efficient working; the loss of one notional job was trivial compared with the substantial and real job loss associated with cost-cutting measures introduced in the 1980s; and the plant indicated a greater degree of security for the site as a whole. But perhaps most important was the fact that single-grade working had been introduced. In effect, the union was prepared to accept the loss of one job in exchange for their long-term aim of achieving this improvement for the average worker. The two changes, then, are closely interlinked.

We have already noted that, partly due to union pressure, management plans gradually modified the traditional hier-archical pattern of working. Along with this, management had recognized that proportionately higher ratios of chargehands to ordinary processmen would be required on the new plant compared with the old (if only because many of the tasks undertaken by the latter were to be automated). The implicit logic of this approach—a movement towards single-grade working—was strengthened and furthered by the commissioning process. The first five operators had already been recruited and trained when the first of a series of delays in construction occurred. It was decided that the operators should stay on the lager plant and assist in commissioning work, a process which would supplement their training by giving them an extensive knowledge of the plant. Further delays meant that these chargehands worked on the plant for nearly a year before normal production began. As a result, a 'team identity' developed among the chargehands, supervisor, and plant manager, as was attested by all three. The manager felt that 'team spirit' was a particularly useful asset and this led him to depart from the manning plans and to adopt the union proposal that only chargehands should work on the plant. Recruitment of ordinary process grades would, it was feared, disrupt team working by requiring the establishment of job demarcations. In addition, single-grade working facilitated the provision of reliefs, a problem that had dominated earlier lager manning discussions.

The grading of the jobs in the lager plant was discussed jointly in the job evaluation committee. The jobs were advertised in the normal way within the company, but union involvement went even further than usual, in so far as the brewery branch chairman discussed the question of suitable candidates with the personnel department. Thus it was agreed that they should all be younger workers and not already on the highest grade (so that they would not suffer a drop in earnings due to a loss of shift premia). Initially it seemed as if insufficient candidates would be forthcoming from the old brewery, but the union pressed the company to readvertise the jobs rather than going outside the site for recruits. This was done, and so all five initial recruits, and all subsequent lager plant workers, came from the old brewery. In addition, the union ensured that traditional training practices were applied to the new plant. Hence they successfully resisted the lager team's wish to give the new recruits less comprehensive training than was eventually provided, and insisted that the practice of operators rather than foremen or managers training new workers be maintained. Management agreed.

The decision to have single-grade, flexible working came about for a variety of reasons. One important factor was the difficulty of applying conventional manning practices to the new plant, given its scale, and the different sets of tasks which workers had to undertake. Another factor of some importance was the union's desire to move towards single-grade working—a system which was adopted in other new sections of the plant. Third, and crucial to the full development of this pattern, was the success of such working when construction delays provided a chance opportunity for experimentation.

PATTERNS OF LABOUR REGULATION AND UNION INFLUENCE

The preceding discussion has shown that the union played some role in shaping the working arrangements relating to the new lager plant. In addition to being informed about management plans at a fairly early stage, albeit in a less than neutral manner, the union was involved in the introduction of single-grade working and was also able to ensure that traditional rules concerning recruitment and training were applied. Compared with the chemical case-study, then, the union in the brewery had considerably more influence. Moreover, little hard bargaining was necessary for the union to achieve this. Much of the negotiation took the form of relatively easygoing, off-the-record discussions, while in other cases the union appeared to play a policing role, ensuring that management conformed to past practice. The assumption on the part of both management and union was typically that arrangements operating at the level of the site as a whole would be equally applicable to the new plant. What all of these factors point to is the way in which the previous pattern of labour regulation shaped working arrangements and surrounding conditions in the new situation. This is not, of course, to suggest that change is impossible: in some situations, one party or another may, for a variety of reasons, seek to introduce major innovations. However, these are likely to be contested unless they also meet with the interests of the other party. In the brewing company change did occur, but it came about in part through chance, and at the same time it accorded with the interests of both management and union. In other respects, labour regulation in the new plant conformed with the general site pattern. Nevertheless, there were important changes occurring at the site, concerning working arrangements and manning levels. These are discussed in the next section. For the present we are concerned with outlining the general pattern of labour regulation. In doing so we look first at those features of the work context which relate most closely to the work process. These include recruitment, promotion, mobility, task allocation, and discipline; we then move on to consider matters relating to pay.

The unions at the site exercised a number of controls over issues relating to work organization. First, the actual number of jobs in each department was fixed in agreements between management and the unions at site level. In addition, individual stewards occasionally raised issues about manning levels, e.g. the failure to fill vacancies, or the balance between overtime and the number employed. Similarly, the unions were involved in wider issues such as subcontracting, and this was enshrined in formal agreements: a proposal to subcontract catering, for example, was rejected by the unions.

Second, the unions were involved in the determination of job tasks. This was most clearly the case in the productivity bargaining which occurred around the time of the decision to construct the lager plant. Before this, such issues tended to be negotiated by the branch chairman and other branch officers at site level. But the scale of the necessary discussions in the productivity deal meant that stewards on the shop-floor negotiated many of the issues involved and had subsequently continued to play a fairly active role in this respect. In addition, the union became involved in disagreements between workers over who should undertake particular tasks and where different sections were in dispute over such issues. In one such case, for example, the branch chairman proposed a compromise between two work groups who were disputing who should undertake an unpleasant task; in another, a disagreement between two shop stewards over task allocation was resolved, not by management, but by a decision of the branch committee.

Third, the union played some role in recruitment. The company had traditionally taken on recruits from members of a youth training scheme, which had been established many years ago. The union had no influence over who was selected for the scheme but did expect that the trainees would be the primary source of new recruits to the company's labour force. In the late 1970s the company proposed abandoning the scheme since there could be no guarantee of permanent employment. The union was opposed to this move and successfully insisted that the training scheme be retained, conceding only that trainees might not automatically get jobs in the brewery.

Recruitment and promotion to particular jobs within the brewery itself were the subject of a formal agreement between the company and the unions' JNC (see below). Under this agreement all vacancies for manual jobs were advertised throughout the brewery and, with certain exceptions, it was open for all workers to apply. Management then selected from among the applicants. The union played no role in the actual process of selection, although—as noted in the previous section—it did discuss the sorts of people who should be recruited to jobs on the lager plant. This, however, was unusual. More generally, there was agreement that applicants for chargehand and supervisory positions had to have passed the relevant promotional tests.

Fourth, the union played a major role in redeployment. In the late 1970s, when work in the old brewery was extensively reorganized under a series of productivity agreements, the TGWU branch chairman discussed details of individual cases with personnel management when redeployment details were being worked out. Shop stewards were also involved in this process, but not the individual workers concerned; they merely retained the right to object to any transfer proposal. Where someone was transferred to a lower-paid job, written agreements provided for the protection of their earnings and pension rights. Workers could also be transferred to other jobs if they were too old or too ill to continue their current work. In such cases they were compensated for lost earnings by an allowance of 90 per cent of the difference in basic pay between the new and old jobs. Transfers of this nature had to be sanctioned by a joint sickness committee.

Fifth, the union played some role in training matters.

Training programmes and procedures were drawn up by the company's training officers, subject to the agreement of the union branch officers and shop stewards. The standards to which operators were trained had to be agreed by the union, and periodical revision of training handbooks was also carried out. This form of union influence was jealously guarded—one instance of this was noted in discussing training arrangements for the lager plant. In another case, the union was able to prevent the work-study manager unilaterally introducing changes in a particular department, and insist that the training handbook be jointly revised by the relevant steward and the departmental manager.

Sixth, general levels of overtime were laid down in site-level agreements, which were modified from time to time. However, the actual operational details were established locally. In the lager department operators decided among themselves who should work overtime, drawing up a rota which they presented to the foreman. They also regulated access to abnormal overtime, for example when someone had to be called in to meet agreed manning levels. In other departments such relief working was arranged by the supervisor, who kept a log of relief arrangements which was open to inspection by workers and stewards to ensure fair play.

Seventh, the union played a significant role in disciplinary matters. Individual workers could be given formal warnings about their conduct by departmental managers, but only in the presence of a shop steward. Where this procedure had not been followed, the union had been able to get the formal warning withdrawn. If a manager felt an offence warranted suspension or dismissal, the case had to be heard by the site disciplinary committee, which consisted of union and personnel department representatives. Both sides of the committee co-operated in establishing the facts, but the final decision rested with management. The union strongly defended anyone it believed was unjustly accused and sought—generally with some success—to mitigate punishments meted out to those who had breached the disciplinary rules. Departmental managers also sometimes gave general warnings to workers concerning their work; stewards would be involved in discussing such warnings and frequently tried to defend the behaviour of their constituents.

In addition, stewards were often involved in a wide range of other matters, e.g. concerning health and safety and working conditions. Health and safety matters were also monitored by health and safety committees and departmental consultative meetings. Issues were usually initiated by individual workers or stewards and were then taken up in the committees. Stewards were typically involved in examining the nature of the problems and the search for a solution. When stewards felt that they were making little progress, they could take the problem to the branch committee or to a central safety liaison committee. Most health and safety issues were, however, settled locally.

The holiday allowance was the subject of a formal agreement, but the detailed allocation of holiday time was determined entirely by the operators themselves. It was also relatively easy for workers to take casual leave, since they could swap shifts with friends as long as the foreman was informed. Working hours and shift times were also laid down through a collective agreement at site level, although one department had negotiated different shift times. In the lager plant, rest and meal breaks were taken by operators largely at their own discretion, subject to the discipline of the group. These arrangements were more formalized in other departments, but were subject to collective agreement.

The unions also negotiated over a variety of pay issues. Basic pay was based upon job grades, which, in turn, were determined by job evaluation. The evaluation process was operated jointly by management and the unions. When the system was introduced, each job on the site was discussed by a panel of four managers and four union representatives, and an agreed profile of each job drawn up. This was used to rank jobs, which were then categorized into grades. The system had been revised several times, as provided for in the first agreement, and on each occasion the union was fully involved. New jobs, such as those in the lager plant, were also discussed within this established framework. Grades were initially allocated to new jobs on a temporary basis and could be challenged after six months by any of the workers concerned. In such cases the job would be fully studied by a joint union–

management team, which would establish an appropriate profile and grade.

Wage negotiations took place annually, and the JNC also used these occasions to raise such matters as longer holidays, shift allowances, and overtime rates. Over the period studied, the unions had achieved, overall, wage rises in excess of the rate of inflation. But they had been less successful in their demands on other issues; the falling demand for its traditional ales made management unwilling to make other concessions. However, the union demands for other benefits were a factor leading to the productivity agreements. These led to some redundancies but also to improved pay and conditions and, it was believed, greater job security in the future. As part of the productivity deals which were struck during the introduction of the lager plant, workers received substantial pay increases and also a continuing share, through a performance and efficiency bonus, of the cost savings achieved.

The union had also negotiated a sickness agreement which provided for relatively generous benefits—sick pay was average earnings. The agreement also set up the sickness committee referred to earlier: it had equal numbers of management and union representatives and met twice a week. The committee reviewed individual sickness and absence records and controlled the payment of sick pay. The union regarded its participation as essential, so that it could protect the scheme for the majority of members from any minority that might try to abuse the provisions of the agreement.

There was also a structure of consultative committees at departmental and site levels. These discussed a wide range of issues, e.g. information on production trends and prospects, plans for and progress with the introduction of new equipment, and production problems. Stewards also used these committees to raise issues on which they had failed to make progress through less formal channels. Generally, these consultative committees were union-based. The exception was the lager plant, where, due to the small number of workers and the pattern of working, it was agreed that any workers who so wished could attend and contribute.

The key body in the consultative structure was at site level. Its membership consisted of senior managers from the main functions and departments, together with other managers brought in to discuss special issues, and representatives from the various union groups on the site. Reports on production in each department were given, along with marketing reports and information about the state of construction projects and capital project plans. It was at this committee, for example, that the unions were first formally told about the plans to construct the lager plant. Occasionally senior managers presented an overview of the company's strategy, as regards both brewing and its other interests. The union side was able to ask questions about these and other aspects of company policy and to seek additional information. The union also occasionally put up its own proposals at this body, e.g. for the construction of improved packaging facilities. It also used the committee as a forum for raising grievances and issues which had not been satisfactorily resolved. The union representatives were thus not just passive recipients of management information.

Strong bargaining relationships between senior managers and the chairman of the brewery branch and other key union officials were also a feature of industrial relations at the site. Such relationships were an important source of information, albeit of a kind which often could not be widely disseminated. The plan to build the lager plant, for example, was known in this way to a handful of union leaders before the information was formally released within the company. In addition, a good deal of bargaining occurred through these informal relationships. As noted previously, much of the discussion over manning arrangements and health and safety issues in the lager plant took place informally between the brewery branch leaders and the managers on the lager project team.

While the unions played only a consultative role on a range of issues concerning general corporate strategy, they did have a considerable influence over both sides of the wage–effort bargain. Hence, it is clear that the role they played over the lager plant was broadly in line with their more general role. Furthermore, the degree of union influence in the brewery is in marked contrast to the part played by the union in the chemical case-study, where, it will be remembered, the degree of joint regulation was very limited over aspects relating to work organization. Beyond this contrast, the forms of control were broadly similar, except that the degree to which task performance was detailed in standard operating procedures or similar sets of rules was much less in the brewery. As in the chemical case-study, however, there was, more generally, a relatively bureaucratic structure, although the rules were more often developed, amended, and applied jointly. The technology played a similar role in shaping work tasks and the level of supervision was broadly comparable, except in the lager plant. The main contrast in labour regulation between the chemical and brewing case-studies, therefore, lay in the fact that there was a much greater degree of joint regulation in the latter. This is to be seen most clearly by considering the history of bargaining over working practices and manning levels.

BARGAINING OVER WORKING PRACTICES, COSTS, AND EFFICIENCY

Just to outline the main issues and the key events in the bargaining processes concerning working practices and related issues would require a chapter in its own right. Given the constraints and primary interest of this volume, we shall therefore confine ourselves to highlighting a number of key features of such negotiations, which, at least in the 1960s and early 1970s, were generally termed 'productivity bargaining'.

Such bargaining was not a new phenomenon in the late 1970s. That is, it is wrong to suggest that negotiations over more efficient methods of working and job reductions were first introduced in the changed labour market and political conditions of the time. Indeed, such bargaining had been a virtually constant feature of industrial relations at the site since the late 1960s. From that time, five major sets of negotiations had occurred. The major impetus, in every case, was the deteriorating sales of the company. This meant that there was pressure, first, to align manning levels to the reductions in output and, second, to seek other economies. Both involved changes in working practices and related issues. However, the precise focus of agreements varied. The earliest agreement was a typical productivity bargain of the period.

The second, in the early 1970s, concerned changes in working practices when there was major capital investment in one section of the plant. The third, which followed on immediately from the second, aimed at cutting staff costs by a fifth. The fourth set of negotiations again began immediately after the previous bargain had been struck and focused on cost-cutting, mainly in non-production areas. Finally, in the early 1980s, a further series of negotiations sought to cut costs and increase efficiency, and included a move to five-day working. There had therefore been virtually continual productivity bargaining for almost fifteen years by the time of our research; the only gap was in the early 1970s.

One reason for this virtually continual process was that deals typically took a long time to complete. Indeed, if anything, the period of negotiations increased. The first two sets of negotiations were each completed in less than two years; the second two each averaged between three and four years; and the final set took almost as long. This was in part attributable to the scale of change proposed in later agreements, but it also indicates the scale of negotiation which was involved. In the most recent bouts of productivity bargaining, management had sought to expedite negotiations by short-circuiting procedures and even appealing directly to the workforce. The unions, with support from the membership, insisted that procedures be followed to the full, even though they were sympathetic to management's concerns.

Through detailed negotiation the unions were able to achieve significant modifications to management's proposals. Here we shall concentrate upon the negotiations which occurred during the period covered by our research—that is, from the mid-1970s—and list the key areas and issues. First, management proposed on a number of occasions that manning should be reduced and work be done by contractors. Such proposals were confined largely to transport and ancillary services such as catering. It should be noted that a good deal of transport work had been subcontracted up to about the mid-1970s; hence, in this area management was proposing a return to past practice rather than a major innovation in meeting its labour requirements. Despite this, the unions were opposed to what they termed 'the major principle of contract workers'; the brewery branch, for example, declared 'we are not prepared to accept cuts and give work to outside contract'. The unions prevented the subcontracting of catering and other ancillary services, albeit accepting changes in working practices and manning levels, in addition to lower standards of food provision. Lorry drivers achieved a commitment from the company that its own lorry fleet would be fully utilized before any use of contractors was made: even then they were to be used only to carry ale from warehouses scattered around the country rather than working from the site itself. The unions also accepted that some contract chauffeurs would be employed.

Second, a great deal of negotiation focused upon manning levels and working arrangements. Management proposals took a number of forms, but two were of especial importance. The first concerned changes in the tasks to be undertaken by different groups. In some cases, they involved the fusion of work groups, or workers assuming tasks which had previously been done by others (some of these proposals also involved the use of new machinery and equipment, e.g. fork-lift trucks for the loading of vehicles). That is, these proposals concerned increasing flexibility. Others, however, effectively meant reduced flexibility, e.g. the suggestion that certain groups of maintenance workers should work solely in specific areas of the site. The unions were generally prepared to consider such changes, with the exception of any weakening of certain craft demarcations. On the latter issue management had made little progress by the end of our research, despite having sought such changes for many years. Elsewhere the unions insisted on detailed negotiations and pilot schemes. As a result, the unions were often able to ensure more satisfactory working arrangements and to reduce the scale of job loss.

From the mid-1970s a central feature of negotiations concerned changes in working hours and shift arrangements. In the case of maintenance workers, for example, management wished to introduce shift-working in order to reduce rota'd overtime and call-in payments. The maintenance unions were prepared to negotiate these issues, but insisted on additional payments on a scale which management was not prepared to agree. As a result the proposals were not implemented. In many production and related areas management wanted to move from seven- to five-day working, in view of output requirements. This was finally achieved after a great deal of negotiation over compensatory payments and related questions of redundancy (these are discussed below). In addition, management wished to make more effective use of labour by removing standard shift-manning levels in a variety of areas and calling in labour as and when it was required (that is, according to the needs of the stage of the production process). The unions opposed this and were able to achieve an agreement on standard manning levels for each shift, although management had the right to call in such additional workers as might be needed for exceptional workloads.

A key issue of discussion which ran through all the changes proposed by management was money. In some cases, as has been noted above, management proposals were rejected largely because additional payments were deemed insufficient. However, on many issues a central question concerned compensation for lower gradings of redeployed workers and loss of overtime earnings or other forms of additional payment. As has been noted in the previous section, management finally made commitments to guarantee earnings. In addition, further payments were made for agreeing to changes in working practices, which, in later stages of the history of productivity bargaining, were further supplemented by self-financing elements.

Given the scale of job loss, it is not surprising that a great deal of discussion centred around alternatives to redundancy, and redundancy payments. It was agreed that any vacancies occurring anywhere in the plant during the period of negotiations would be left unfilled to facilitate redeployment, and management also agreed to provide some help in job creation. In addition, the unions sought to improve upon management's proposals concerning the level of redundancy pay. While, in general terms, the company was not prepared to move beyond its principle of offering only double the statutory rates, the unions were able to achieve detailed changes which had the effect of improving the payments received by many of those finally made redundant.

Finally, an important area of union activity concerned

limiting management discretion and individualistic approaches. The unions successfully insisted upon a number of points, which included the following: redundancy payments would be no greater for volunteers than for those made compulsorily redundant; no 'private deals' between management and workers would be permitted; the criteria for the selection of those to be made compulsorily redundant should be confined primarily to length of service (thus preventing management from selecting workers on the basis of assessed ability etc.); early retirement was to be favoured; volunteers for redundancy should be accepted from anywhere on the site; and redeployment options should similarly be site-wide. In addition, they ensured that changes negotiated were acceptable to the groups most immediately concerned. The unions were able to win agreement on these issues despite the fact that, in the latest round of productivity bargaining, the number of volunteers for redundancy exceeded the number of jobs to be cut.

Through the productivity deals, management did achieve a substantial reduction in manning levels. In the six years prior to our research, for example, total employment (including staff) fell by almost a quarter on the site. But two points should be noted: first, the proportionate fall in employment was broadly in line with that in output; second, there was no significant increase in the rate of job loss from 1979/80. Those who remained in employment received substantial increases in earnings, as well as compensation for lost overtime etc. More generally, the unions succeeded in changing management's proposals to a significant degree. However, what should also be stressed is that the unions accepted in principle the need to cut costs and, indeed, as sales continued to decline, became increasingly prepared to co-operate with management. But such co-operation was conditional. First, they were not prepared to accept any bypassing of the unions or short-circuiting of procedures. Second, they wished to ensure that the scale of change was limited to what they saw as necessary. Third, they tried to ensure that workers were adequately compensated for any changes in their situation. Indeed, on a number of occasions they threatened to withdraw from discussions unless management improved their offer. Furthermore, in an attempt to provide greater job security the unions put forward the proposal that a canning plant be constructed on the site; this was rejected by management. In short, the unions played a central role in the changes which occurred. Despite adverse market conditions, the union was able to win significant changes to management's proposals and, in some cases, to prevent any change at all. Management, therefore, achieved only a good deal of what it sought, but it did so under tight conditions laid down by the union. Such union influence depended in large part upon its degree of sophistication, the theme to which we now turn.

UNION ORGANIZATION

Intra-union sophistication and external integration

As far as trade unions are concerned, our focus in this chapter has so far been upon the union which represented production workers. This was the TGWU, which had a site-based branch covering production workers only. In this section we consider the organization of this branch, and then go on to look at union organization more generally at the site, that is, inter-union sophistication.

The TGWU brewery branch had between 450 and 500 members in the late 1970s, when the lager plant was being constructed. The union had a post-entry closed-shop agreement, so that union density was 100 per cent. The branch was organized on two partially overlapping principles —the branch and a network of shop stewards.

In discussing the chemical union organization, we stressed the importance of gaps in steward representation. In the brewery, no such gaps existed. All steward positions were filled, with an average of a little over 20 members per steward. In addition, considerable care was taken by the branch to ensure an adequate representation of different groups and interests in the shop steward body. Hence, for example, the number of shop steward constituencies was increased from 21 to 23 in the late 1970s, to accommodate changes in working practices and shift organization. The initiative for this reorganization came from the branch committee. There had been relatively little shop steward turnover until the late 1970s, when a number of stewards took redundancy. Stewards were elected every two years.

There was also a variety of means by which the union organization was co-ordinated. A key body was the branch committee, which met at least once a month, always before branch meetings, and more frequently during crucial negotiations. The committee consisted of a chairman, three other officers, and four other members. Care was taken to ensure that the membership of the committee was broadly representative of the different areas of the brewery. The committee members were elected by secret ballot every two years, the elections alternating with those for shop stewards. The chairman and vice-chairman had both held office for over ten years, but several of the other members had been on the committee for only a few years. This reflected turnover due in part to redundancies and in part to the general pattern of promotions and retirements. Shop stewards accounted for about half the membership of the committee. The minutes of the committee reveal that meetings were well attended. In addition to dealing with general union matters largely external to the site, the branch committee spent considerable time discussing work-based issues. Indeed, it can be seen as the key body in which general union policy was formulated. The committee in effect acted as a group of senior shop stewards.

The second important form of co-ordination was the branch chairman. He worked full-time on union business and was paid by the company. As well as dealing with routine branch matters, he played a key role in all major negotiations, not only those for the brewery branch itself but also those involving the other unions and branches on the site (see below). He was also the key source of day-to-day help and advice for other shop stewards and had a network of strong bargaining relationships with a wide range of managers.

The third form of co-ordination was the shop stewards' committee. This was integrated into the branch structure, both through an overlap of membership and also because the branch chairman chaired the committee. It functioned primarily as a communication channel to and from the branch committee. Meetings were less formal than those of the branch committee and until the late 1970s had been held irregularly. Since then, as noted in the previous section, shop stewards had played a more important and active role, and this was reflected in a greater regularity of steward meetings.

The branch meeting was the fourth form of co-ordination. Unlike the meetings of stewards and the branch committee, these meetings were held out of working time and were usually attended by 10 per cent or less of the members (this was a considerably higher attendance rate than in the chemical union). Most shop stewards attended branch meetings regularly. Special meetings of the branch could be held in working time, and these were usually attended by 300 or more members (about two-thirds of the total). Branch meetings were usually concerned with reporting on the state of negotiations and various site-related issues, as well as general union information such as national campaigns. Proposals from the branch committee, e.g. in relation to the annual negotiations over pay, were put to the branch meeting. In this way the branch ratified the principles to be followed by its negotiators. The branch also made decisions in cases where there were serious disagreements between different work groups or stewards. The advice of the branch committee was generally followed, but this was not invariably so. On a number of issues, particularly those which were not primarily confined to site matters, the branch had rejected the recommendations of the branch committee.

Thus the brewery branch was sophisticated. It had total membership in its area, bolstered by a closed-shop agreement. The pattern of steward constituencies covered the key groups and different interests, and all steward positions were filled. In addition, it had four mutually supporting means of co-ordination—the full-time branch chairman, the branch committee, the branch meeting, and the shop stewards' committee.

Moreover, the branch was relatively well integrated with the larger union. Regional and national officials very rarely came on to the site and were not involved in negotiations, even over the annual pay increase. Rather, the integration took the form of key union officials from the branch being involved in various committees of the larger union. These included national and regional trades advisory committees relating to the brewing industry, and a national committee on the food, drink, and tobacco industry. The committees primarily fulfilled an information exchange function and did not attempt to impose policies upon branches or workplaces. Branch activists identified closely with the larger union.

The introduction of the lager plant had little effect upon the general pattern of organization in the brewery branch. In the reorganization of shop steward constituencies, one was allocated exclusively to the lager plant. This was decided by the branch committee, despite the small numbers employed on the plant, since it was deemed important that the special interests of these workers should be represented. The lager steward played a somewhat less active role than many of his counterparts on other parts of the site. This was due to the small number of workers on the plant and the pattern of working. These meant first, that workers had more opportunity to sort out problems among themselves, rather than seeking the support of a steward in raising issues with a foreman. Second, the informality of the plant facilitated direct discussions between workers and supervisor where these were felt to be necessary. Third, as noted previously, it had been agreed that all workers could attend the lager plant consultative committee, and this meant that the steward did not have a monopoly over this form of communication and discussion. The reduced role of the union, however, was of limited significance and certainly did not result from any concerted or conscious strategy on the part of management.

Inter-union sophistication

Not only was the brewery branch organization sophisticated, but so was the inter-union organization among the manual unions. However, the brewery branch had little contact with the two in-house staff associations, one of which represented clerical and managerial staff, the other foremen. Some foremen were also members of the TGWU, but the union had no negotiating rights on their behalf. Contacts between the staff associations and the brewery branch were confined to occasional discussions concerning such issues as pensions and the restructuring of the company; even these contacts were of recent origin, being largely associated with negotiations over job reductions and greater efficiency.

All the unions representing manual workers had 100 per cent membership in their areas and were supported by closed-shop arrangements. The TGWU was responsible for organizing the majority of manual workers on the site. In addition to the brewery branch, there was also a transport branch of about 200 members, which covered the lorry drivers and others working in the tanker station. The transport branch did not allow stewards to sit on the branch committee, although they could and did attend and speak at its meetings. The stewards formed their own shop stewards' committee.

There were between 150 and 200 engineering and craft maintenance workers on the site. As is typical of craft unions, they belonged to a variety of branches based on place of residence. On the site, therefore, the basis of organization of these groups was the shop steward constituency. Electricians and instrument technicians, who together constituted the largest group of tradesmen, belonged to the EETPU, and were organized into two steward constituencies, one for the power station, the other for maintenance electricians and instrument technicians. Maintenance fitters belonged to the AUEW and were also represented by two stewards; they formed the second largest group of craft workers. Carpenters and painters were members of UCATT and each had their own steward. Finally, there was a small number of coppersmiths, who belonged to the NUSMWCHDE and also had their own shop steward. The seven trades' shop stewards jointly formed the trades' shop stewards' committee, which elected a convenor who was not a steward as its chairman. The structure of representation of the craft workers had remained unchanged since the early 1970s. So far as site and company issues were concerned, the various trade groups operated independently of their larger unions, whose full-time officials rarely became involved in issues concerning the company.

There was a very high level of co-operation and co-ordination between the various union groups on the site. Some twenty years previously bargaining arrangements had been changed to provide a single bargaining unit covering all manual workers. The unions established a JNC and a Trades Union Advisory Committee (TUAC), in which all the unions on the site were represented, to discuss union policy and agree negotiating positions *vis-à-vis* the company. The JNC was made up of the chairmen of the two TGWU branches and the convenor of the Trades Group. These three men also sat on the TUAC along with other group and branch officers and senior shop stewards. The JNC and TUAC represented not only union groups on the site but also union representatives at another of the company's brewing sites who were members of the same bargaining unit.

The TUAC met regularly every month, and in addition held six or more additional meetings per year to deal with wage negotiations and urgent or complex matters, such as changes in working practices. The TUAC sought to establish a common position on issues facing the unions, and was thus the body where differences between the groups were reconciled before negotiations with the company. The TUAC was therefore a policy-making body. Its decisions were reached by consensus, thereby precluding domination by the larger union groups involved. However, a key principle in its deliberations was that the interests of the workforce as a whole should prevail over sectional interests. The consensus rule and the perspective to which the committee was in principle committed meant that discussions were often protracted. Decision-making was further lengthened by the requirement that policies initiated or significantly modified by the committee had to be ratified by the various union groups. This, however, did sometimes provide a means of resolving deadlocks as members changed the mandates to their representatives. Alternatively, the JNC might seek concessions from management in order to give more scope for the different union interests and demands to be accommodated. The JNC as such met only to conduct formal negotiations with the company.

Co-operation among the various union groups certainly involved considerable strains at times; reference has already been made, for example, to differences of view over the negotiation of productivity deals. However, these were generally resolved and in this way a significant degree of unity maintained. Thus, the unions were typically able to present a common front to management. At the same time, this high level of inter-union sophistication provided a structure and a commitment to the general interests of the workforce, and thereby to a broader perspective on the part of the unions than might otherwise have been the case. This approach was strengthened in the case of the brewery branch by the nature and size of its membership and its own internal sophistication. This broader perspective led the unions to agree to productivity deals and associated reductions in the labour force. It was accepted that efficiency had to be increased if the site was to continue in operation. But strict conditions were imposed by the unions, and these led to important modifications to management proposals concerning work organization, along with significant improvements and guarantees concerning pay. The nature of the union organization permitted it to achieve significant controls over detailed aspects of work organization. These controls were maintained in relation to the lager plant, and the brewery branch saw considerable movement towards its goals of improved gradings and changes in working practices.

In both the chemical and brewing cases, market considerations were important in the introduction of new technology. An associated consideration in the former case, however, was improving quality and yields; this, it was believed, could best be achieved by transferring many aspects of control from the workforce to the technology. In the brewing case, labour considerations centred initially upon the site more generally— avoiding stoppages of work and using the plant as an inducement to productivity deals. The details of work organization evolved in very different directions in the two cases—in chemicals towards a polarization of skills, in brewing towards single-grade working—despite the basic similarities of the core production process and the new technology. While union influence cannot be seen as the major factor, it is nevertheless true that in the brewing case the union did play a role in this change, as well as in other aspects concerning the new plant. In chemicals the union played a negligible role. In both cases the pattern of union influence in relation to the new technology reflected the pattern of joint regulation more generally. The actual forms of labour control in the two cases were quite similar, but in the brewery the role of the union served to change the effects and mode of application of these various forms. For example, the brewery workers were far more collectively orientated because of union involvement in the formulation and application of rules; structures were biased towards collectivism. Hence, for example, the existence of an internal labour market which was very similar to that at the chemical site did not have the same effects upon worker attitudes and behaviour. In the chemical case, however, considerations relating to the internal labour market appeared to shape workers' perceptions of the role of stewards. This was not so in the brewery where there were collective means to resolve grievances and dissatisfactions which were generally of greater significance. Their existence reflected and reaffirmed the degree of inter- and intra-union sophistication.

CRAFT ADMINISTRATION AND CRAFT UNITY: THE CASE OF SMALL-BATCH ENGINEERING

In this chapter we turn to a quite different type of production process from that considered in the preceding two chapters— small-batch production in engineering. The work in the spares plant studied was of a craft nature and could have been endangered by the introduction of CNC (computer numerical control) machines over a number of years from the late 1970s. The first section of this chapter looks at the reasons why management introduced this new equipment, and goes on to outline the way in which the work undertaken by operators on the CNCs compared with that of their counterparts on conventional machines. There were a number of significant changes relating to the need for programming and the fact that some of the new machines undertook a wider range of functions. However, traditional craft skills were still necessary, although they had in part to be used in new ways, for the principle of full operator programming became established gradually. The second section of the chapter goes on to look at how this and other labour-related decisions came about. An important factor running through the gradual evolution of management thought was the importance of the craft ethos for the continued viability of the plant. They therefore finally chose to use the new technology to strengthen this element, rather than weakening it through removing tasks from the shop-floor, and instituting a more hierarchical and bureaucratic system of production organization. The union, however, also played a role in this and many other issues concerning the new equipment, and sought, successfully in the end, to establish a new technology agreement and to ensure that all workers won some gains from the use of CNC machines. The significant part played by the union in relation to these questions reflected its more general role in the pattern of labour regulation. This is discussed in the third section. A final section relates the degree of joint regulation to the pattern of workplace union organization.

CNC MACHINES AND CHANGES IN WORK ORGANIZATION

Production in the spares plant was for small numbers of parts (batches of more than thirty were rare) of a light engineering character. All production equipment was laid out on a single factory floor, with functional groupings of machines into shops separated by open gangways. The milling shop was the largest, and consisted of equal numbers of horizontal and vertical milling machines. The turning shop was only slightly smaller, and was characterized by a proportionately large number of centre lathes, and relatively few capstan lathes. These two shops accounted for more than 60 per cent of the production workforce, the rest being distributed between fitting, press, grinding, assembly, and jig-boring shops. This section will be confined to a discussion of the milling and turning shops, since it was only in these that CNC machines had been introduced. In the other shops, machines were controlled conventionally—their operation required manual adjustment and setting, and manual tool changes. Some machines had had digital readout and recording systems attached, which facilitated accurate setting and permitted a limited degree of automatic sequencing.

A total of nine machining centres and three CNC lathes were installed in the early 1980s. The machining centres were placed in the milling shop and the lathes in the turning shop, but the two types of CNC machine tool were set along the gangway separating the two main shops so that they were all sited together. All the machines had manual data-input systems, but the precise nature of these varied. Two of the machining centres had only limited programming facilities on the machine, and consequently the bulk of their programs had to be prepared elsewhere. Three more machining centres had more sophisticated computing facilities, but also required additional and separate computing power for complex parts. The remaining machining centre and the lathes all had manual data-input systems which had been specifically designed for operator programming and were of a 'menu' type, where the operator had to choose between various options displayed on a small VDU. No additional computing facilities were needed for the lathes, but as this machining centre was still being commissioned when the study took place, it was not known to what extent programming might have to take place off, as well as on, the machine.

The CNC machines also differed with respect to the type of operating programs with which they came equipped. The 'menu'-type machines had programs which, for example, set limits to the speeds at which particular metal types could be machined, as well as covering features common to any machining activity (such as the limits within which the tools could move). These programs could, of course, be altered, but did not need modification in the normal course of machining. Other machines, particularly the three machining centres with more complex manual data-input facilities, had minimal operating programs. Each time a program was written for machining a part, for example, instructions had to be made to set the parameters within which the machining table could move, and even to ensure that the coolant pump was switched on and off. The nature of 'normal' programming thus varied from one group of CNC machines to another.

The introduction of CNC machines

The company was among the first in Britain to use NC and later CNC machines in its main factories, but did not seriously consider them in the spares plant until late in the 1970s. One machining centre was bought and then, a couple of months later, the senior engineer circulated a memorandum suggesting that more machines would be needed in view of the potential workload that had already been identified. Three sets of considerations were important in shaping this proposal. First, crude comparisons of the time taken to produce parts on conventional and CNC machines indicated that substantial savings were possible. A potential advantage of CNC machining centres for the small-batch needs of the plant was that total metal-cutting time could be reduced. A part might, for example, require drilling and several different milling operations—all of which could be performed on a machining centre with only one setting. With conventional machines the part would have to be sent from one machine to another to have the different operations performed. All the transfer time, as well as that needed to reset the part in each conventional machine, was eliminated with the machining centres. Shorter production times would reduce the time a customer had to wait for an order, thus strengthening the company's service strategy. Secondly—and this in part explains the timing of the decision to introduce CNCs—the plant was increasingly required to make parts which had originally been designed to be made on CNC machines in other plants. These proved difficult to make using conventional machines, and hence, if the production of spares was to be continued, there was a need to introduce CNCs. In addition, it seems that spares managers wished to gain experience of such machines as part of the wider company strategy of developing the spares service.

The chief engineer proposed purchasing five more CNC machines, three of which would be installed in the main spares factory, and two more in another plant. Initially these proposals were accepted and acted upon, for two more machining centres were installed in the main factory during the following year. However, the company's overall plans for the introduction of computer-controlled machines were radically revised in 1981. These new ideas were framed within a three-year strategic plan, aimed at improving productivity on the shop-floor, as well as administration generally within the division. The plan also sought to revitalize management. It envisaged that CNC machines would achieve productivity levels 100–200 per cent above those of conventional machines —estimates based upon the crude measurements taken from the first 'experimental' machining centre.

The three-year plan envisaged doubling the number of machines by the end of 1981, and again in 1982, so that by 1983/4 there would be fifteen machines on the shop-floor and several computer-aided inspection and measuring devices. However, it will be clear from the description of the new technology given earlier that these plans were not carried out in their entirety—only nine CNC machines had been introduced by early 1984. One factor which partly serves to explain why the three-year plan was not fully implemented was that over-optimistic projections of CNC productivity potential had been made. Instead of the projected 100–200 per cent productivity gains over conventional machines, management had been able to achieve only a 42 per cent increase on average. Senior managers attributed this to the requirements of small-batch production and the wide range of parts the plant produced. No immediate solution was available, and instead of continuing the CNC purchases management turned their attention to the possibilities of automating the loading and unloading of machines.

As indicated earlier, the CNC machining centres in particular required other items of equipment to facilitate computer-aided programming. A computer time-share contract was entered into in 1981, just before the second machining centre was brought on to the site. It was in use for several years but was ended on the grounds of its excessive cost. However, as we shall see below, this development must be seen in the context of a struggle between shop-floor workers and the production planning department over programming. Several mini-computers, including a portable machine, had been bought by the time our study began.

Work organization on conventional machines

When an order for a part was received it was sent to the production planning department. There the part plans were obtained, and job operation cards were made out, containing brief task instructions and giving the setting and operating times allowable. These documents were put together with progress chasing and record cards into a 'shop pack'. The shop packs then went to the issue department, from where they were sent to the materials cut-off section on the shop-floor. Some job routing to particular types of machine was done in the production planning department, but in the main little work of this kind was done there.

The shop packs and pieces of metal to be worked on were collected from the materials cut-off section by the rack clerk, based in the shop-floor offices of the two main machining shops. The rack clerks, and sometimes the assistant foremen, sorted the shop packs in order to separate urgent orders (working from a list of urgent categories or other instructions from the factory manager) and then put them into machine categories. Occasionally they tried to group jobs requiring similar set-up arrangements together, but more often the variety of jobs involved, together with the pressure to get work done as rapidly as possible, meant that jobs were issued to machine operators largely at random.

Jobs were issued to operators by an assistant foreman, either in response to requests for more work or by prior assessment of operators' needs. Some assistant foremen, for example, went round to see what different operators' requirements were likely to be, and issued work to them in advance of requests. In other areas this was done less often. This method of allocation suggests that the machine operators had little influence over which jobs they would have to do. This was certainly true in so far as difficult work would usually be given to the more skilled men, even though such jobs might make it more difficult for those people to earn bonuses. Operators' freedom of manoeuvre had also been curtailed in recent years by the issue office 'starving' the shop-floor of work. Previously operators had been given up to a week's work at a time, but subsequently they only had at most a couple of days' jobs in hand. However, operators were sometimes able to alter the sequence of jobs to their own advantage by grouping similar types of job together, thus avoiding the need to reset their machines entirely. Some operators worked in small informal teams, sharing the jobs between them on an informal basis, and they consequently had more scope for varying the order in which jobs were performed. The nature of individual operators' relations with the shop office staff could also affect the types of jobs they received.

All the machines in the factory were operated by setter-operators. They were responsible for everything from receipt of the shop pack to handing the part on for further machining, or to the inspection department. The instructions on the job operation cards prepared by the planning department only specified, for example, that they had to mill to a certain depth, or drill holes of a certain size. The drawings in the shop pack were of the final part; they did not specify the operations required on each machine—and so one of the operators' first tasks was to relate the drawings, the job instructions, and the part supplied for working on. The drawings did specify tolerances and qualities of finish which had to be obtained, all of which were very fine as production was to 'tool-room standards'.

Once the operator had received the shop pack, checked that the drawings and instructions referred to the same job, and decided how he was going to tackle the job, he had to obtain the requisite tools and fixtures. These might already be in his possession, or he might have to borrow them from other operators, or collect them from the tool-room or shop office. When this had been done, he set the machine up and cut the part to the specification laid down. Operators usually took the first job from a batch to the inspection department to ensure that the settings were correct, but this was entirely at their discretion. When a job had been completed the operator relinquished control over it by putting it on the inspection rack, although some operators took jobs directly to the inspectors.

The extent of operators' responsibility for the factory's products indicates that they exercised considerable discretion. This conclusion is supported by the limited role that shop-floor supervisors played. In both the turning and the milling shops there were one foreman and three assistant foremen. The assistant foremen were formally in charge of groups of machines on the shop-floor, and so were in a position to exercise the greatest degree of surveillance. However, their main tasks were responding to operators' technical queries by giving advice, ensuring that operators had sufficient work in hand, and dealing with local industrial relations issues. They did not exercise the role of taskmaster to any extent, except perhaps occasionally towards the end of a week, reminding people that they were supposed to be working on their machines. Control over machine operators' working practices was exercised less through supervision than through the combination of limiting the number of jobs they could normally have in hand and the piecework system.

Machine operators' pay could be increased by up to 25 per cent of the basic rate through bonus payments. Operators established a claim to basic pay and bonus by handing in record cards detached from the job description card towards the end of each week. These cards indicated individual operators' output in terms of time units, and were used to calculate individual performance figures. Attainment of minimum performance levels qualified an operator for basic pay, while additional points (up to a fixed ceiling) constituted a claim for bonus pay at so much per point. Performance levels were averaged out over 13-week periods, in order to minimize fluctuations in earnings. Basic pay and bonus earnings thus depended on the levels at which minimum and maximum performance were set, and the value of each performance point above the minimum. Before looking at how these were established, we need to consider the performance calculations themselves.

Performance figures were calculated by the following formula:

$$\left(\frac{\text{time allowed} + \text{allocated time} + \text{small-batch allowance}}{\text{time taken} - \text{'green card time'} - \text{'pink card time'}} \right) \times \text{shop constant} \times 100$$

'Time allowed' was the sum of times allocated for job preparation and operation, the operating time being multiplied by the batch size. 'Allocated time' was time allowed which had been credited to an operator for jobs which, for example, were held up for some reason, although they had been completed. The 'small-batch allowance' was an additional percentage of the preparation time (for setting the machine up) credited to the operator for batches of between 1 and 10.

The sum of time allowed, allocated time, and small-batch allowance was first divided by the length of time an operator could have been operating the machine. This was calculated by subtracting time spent on preparing tools and fixtures ('green card time') or attending quality control or union meetings ('pink card time') from the total clocked-in hours ('time taken'). The result was then multiplied by a shop constant. This was a figure allocated to shops and to groups of machines within shops, as a statistical device to equalize the potential to earn bonus between different types of machine.

These elements of the performance calculation, as well as the levels of minimum and maximum performance and the value of bonus performance points, were all determined through bargaining and negotiation.

The 'time allowed' for jobs had been established by planning departments in different factories where the original parts were made, at various times and using varying criteria. If an operator received a job for which he thought the preparation or operating times given were unreasonable, then he could seek to get them changed. The procedure in such cases was for an operator to return the job card to the supervisor and ask for a rate check. This itself opened up the possibility of bargaining, since the supervisor might decide to agree with the operator's estimate of a realistic time, and enter that on the bonus card. In one shop this procedure was regularly followed with certain categories of work, for rate-checking would have taken up too much time. At other times, and in other shops, such a bargain was struck because the job was wanted urgently. When seeking a rate check an operator usually made the part, but kept it at his machine until a rate check was conceded or agreement was reached between the supervisor and the shop steward for its release. Supervisors tended not to want urgent jobs held up, and so often conceded to demands for an improved rate.

If no agreement was concluded at this stage, then the supervisor sent the job card to the planning department for a full rerating. No physical checks were done in the plant, and the planning department estimated the appropriate time according to synthetic standards. The operator or shop steward had the right to go through these calculations if they still disagreed with the outcome, but were bound to accept it if no discrepancies could be found. The records were altered to give the new rate—although operators also kept their own books of time rates to maintain a check.

If the job times were acceptable, but there were no tools or fixtures available, then the operator had another opportunity to enhance the total amount of time allowed. Such jobs were categorized as requiring 'temporary methods'. Formally the times for the extra work involved in setting up to make such parts could be obtained from the planning office. However, this system led to delays, and there was an informal agreement in one shop that the supervisor would enter an agreed time on the bonus cards. This agreement was subsequently formalized across the whole plant. The amount of allocated time an operator was credited with also depended entirely on local bargaining, although this was less over the amount than when it would be allocated. The small-batch allowances were fixed and not subject to modification through local bargaining.

The total 'time allowed' credited to an operator could be affected by other bargaining relationships. If an operator was asked to stop one job and set up for an urgent one, the operator could bargain with the supervisor over the amount of resetting time to be awarded when the first job was resumed. Previously this too had to be verified by the planning department, but subsequently it was decided on the shop-floor. There were also departmental variations in local bargaining opportunities, which could be used to enhance the levels of time allowed.

The total 'time taken' could also be affected by bargaining, particularly over the time used to make any tools or fittings that might be needed. The length of time spent in various meetings was less open to negotiation. However, the greater this time, and that spent on making tools and fixtures, the smaller the potential metal-cutting time available. Local bargaining was not possible over the 'shop constant' figures which had been set many years ago. Constants had been altered only when it was very evident that they seriously affected a group's chances of making bonuses.

It is thus apparent that most of the performance calculation elements were affected by local bargaining, mainly between individual operators, or shop stewards, and supervisors (the system and practice was therefore very similar to that found in much of the engineering industry: see e.g. Brown 1973; NBPI 1967). The performance levels and point values were also set through bargaining during the annual wage negotiations. Since the factory opened, both the minimum and maximum performance levels had been raised, as had the value of performance points.

One relatively unusual feature of the bonus system in this

factory (although reminiscent of the origins of the Coventry 'gang system': see Rayton 1972) was that it applied not only to individuals but also to voluntarily formed groups. In some shops, groups of up to six or seven workers had combined to form teams, which shared work and the bonus earnings equally among themselves. These informal groups were recognized by the company so far as pay was concerned. Elsewhere individual bonuses had been replaced completely by shop bonuses, though this had not happened in the turning or milling shops. In all cases these moves were made at the operators' initiative.

CNC machines and work organization

Although the CNC machining centres and lathes were placed in the milling and turning shops respectively, they were all located together. This permitted an assistant supervisor in the milling shop to deal only with the machining centres and the shop-floor mini-computers. The assistant supervisor responsible for the CNC lathes in the turning shop, however, also had other duties. This difference partly reflected the smaller number of CNC lathes compared with machining centres, as well as the fact that the latter required computer aids for complex programming.

The CNC machine operators were not paid piecework but had guaranteed earnings. In practice this meant they earned the equivalent of full bonus on top of basic earnings. This system was introduced because CNC jobs were not independently time-rated from the start, and so bonus earnings could not be calculated in the normal way. Although this meant that there was no economic compulsion on the CNC operators to work hard, and—as we shall see— supervision was in some senses even less direct than in conventional machine areas, they were subjected to several other forms of pressure to achieve output. First, the managing director made a practice of walking round the factory at least once a day at different times, and paid particular attention to the use being made of the CNC machines. His attention extended to asking operators why their machines were not cutting metal, or why there had apparently been no progress since the previous day. CNC operators were also under pressure from conventional machine operators to be seen to be working all the time. The absence of piecework meant that they could less easily vary their work-rate throughout the week.

All the operators did their own programming on the machines, and had thus taken on the task of interpreting drawings into program terms before machining operations could begin. However, as indicated earlier, the implications of taking up this task varied from one machine to another, depending upon the type of programming possible. In fact there were quite extensive changes in work and work organization in the machining centre section—more so than on the CNC lathes. Before looking at this in more detail we should note that other machining tasks, such as setting tools and parts and preparing fixtures, did not differ significantly from work on conventional machines, or between the CNC lathes and machining centres.

Jobs for the machining centres were sometimes selected by the planning department, which routed jobs directly to that section of the milling shop. The effect of such control was mitigated by the fact that the machining centre supervisor sometimes put jobs back on to conventional machines, either because they were not in his view particularly suitable for the CNCs, or because the machining centres had become overloaded. In the latter case, if work was kept until there was time to do it, the aim of providing a rapid customer service would have been endangered. The machining centre supervisor also took jobs which the milling office indicated might be suitable, and even went round other parts of the shop selecting jobs from those sent in from the materials cut-off section.

Allocation of jobs to individual operators in the machining centre section followed much the same pattern as in conventional areas, although the operators themselves often selected the next job instead of waiting for the supervisor to give it to them. Thereafter, however, work was significantly different from that done on the conventional milling shop machines. First, the tooling used on the machining centres differed from that on conventional milling machines, partly because they were CNC machines, but mainly because they were capable of performing more machining operations. The CNC machining centres could perform all types of milling operation and, in addition, could be used to drill, ream, and tap jobs. This meant that the operators, who all came from the milling shop, had had to learn how to combine milling with other machining operations. At the same time they had had to learn how to program such operations in the most appropriate sequences, in order to avoid resetting the job after performing certain operations. Operators had to conceptualize all the operations and devise the best sequence for performing them. In other words, machining centre operator programmers had taken on production planning tasks in order to be able to program effectively. For programming itself operators had had to learn codes appropriate to their particular machines (the more complex machines had about forty different entry codes), and they had to be able to use the two shop-floor mini-computers. In practice, however, only two operators were able to use the additional computing facilities; the others had not yet been trained at the time of our study. The new machining centre with the 'menu'-type programming system required less extensive programming knowledge, but did permit the entry of new programs while a machining program was actually running. The other machines would take only one program at a time; once these had been programmed and set up there was nothing for the operators to do except monitor their performance. One operator, who was particularly skilled at programming, utilized some of the 'waiting time' while jobs were running on one machine to write programs on the shop-floor mini-computer.

CNC operators on both the machining centres and the lathes filled out operating records for each job, showing the time taken in programming, setting up, and operating. Machining centre operators also wrote up fixture and setting notes for each job. These notes, the job program, and the fixtures were stored on the shop-floor; one copy of the program was sent to the planning department for safekeeping. Programs for the CNC lathes, however, were not recorded, since it had been found easier and quicker to reprogram each time a job was done. When jobs came up which had already been done on a machining centre (and which had been automatically routed there, since the job card would have been altered) operators could simply look out the tape and machining notes, and go to the 'library' to select the appropriate fixture. However, there were no hard and fast rules on this. Operators might program afresh or alter earlier programs and fixtures. In fact, the variety of parts made in the factory meant that, at the time of the research at least, there was little utilization of previously recorded programs.

One consequence of the way work was performed on the machining centres was that the role of the supervisor had changed. The assistant supervisor in charge of the machining centres had previously looked after a section of the milling shop. Thus, first, he had no experience of supervising other machining tasks, such as drilling or reaming, which could be performed on the CNC machines. Second and more important was that he had no experience of working on CNC machining centres, and had only just started using the shop-floor mini-computers. He was therefore in no position to offer the kind of technical advice that assistant supervisors normally gave. Indeed he generally relied on the expertise of one or two operators when it came to programming problems. Similarly, operators assisted each other to a greater extent than normal when machining or programming problems arose. The operators were to a great extent a self-supervising team, while the assistant supervisor was primarily concerned with the administration of the section.

Operation of the CNC lathes did not lead to such large changes in the turning shop, mainly because programming was simpler. All the CNC lathes had 'menu'-type programming systems, operating on a question and answer basis. Set-up and operation of the new lathes were very similar to work on the conventional machines, except that there seemed to be more waiting about while programs were running. Tooling was simplified with CNC lathes, and the range of work performed on them similar to that on conventional machines. Here also the supervisor's role was effectively reduced, since he had no experience of working with the new machines, and so operators solved problems among themselves.

Despite the greater productivity of the CNC machines, so that fewer workers were, in principle, required to produce a given output, there was no actual change in the level of employment within the plant itself. The total number of jobs on the shop-floor did not change in the early 1980s, and when somone left he was replaced. Jobs were certainly transferred to the CNC machines, and fitters in particular had less work as a result of the accuracy of the new machines. However, it was impossible to separate the effects of these changes from the impact of general shifts in the nature of customer orders. The question of potential job losses within the plant had in any case been met by the company's plan, made at the same time as the extensive CNC plans were outlined, to retain the same number of jobs overall during the three-year plan. Thus, where workloads had fallen workers were transferred to other departments, particularly the assembly shop, where operators were needed to work on a developing line—the reconditioning of old machines. Some jobs were lost, however, in factories which had previously done work under contract for the company, since more work was done in-house with the introduction of CNC machines. The stability of employment in the plant reflected the growing importance of servicing and spares as sales of new machines declined. In the company as a whole, the latter had led to substantial job losses and plant closures over the period covered by our research. One final point is also worthy of note. As will be seen in the next section, the union was concerned that the introduction of CNC machines would adversely affect the bonus earnings of conventional operators. In fact, the reverse of this occurred. A detailed analysis of piecework data shows that effort declined relative to 'performance' with the introduction of CNC machines. This occurred because larger-batch jobs were diverted to the CNC machines, with the result that conventional operators were more likely to be paid small-batch allowances (Gourlay 1986).

LABOUR FACTORS AND THE UNION'S RESPONSE

The company formally announced its intention to introduce CNC equipment to the union in the course of the annual wage negotiations, immediately after it had made the decision. This led to two sets of reactions from the union. The first of these was to demand a new technology agreement. This, however, had only a small effect upon the way the new machines were introduced and worked, since agreement was not reached until after they had been installed and were working. Hence, we need to look separately at decisions relating to working arrangements on the new machines, the focus of the second union reaction. We look first, however, at the question of the new technology agreement.

The New Technology Agreement

The company's formal announcement of its intention to introduce CNC machines led shop stewards to propose seeking a new technology agreement. This was a rather unusual step for the AUEW, but was taken for a variety of reasons. Shop stewards in the plant felt that new technology was a distinctive issue, compared with past changes. Central was the question of operator programming (we discuss this at length in the next section). But, in addition, the introduction of CNCs was seen to pose the need for some form of agreement covering job security, training, safety, and the pay of CNC operators. Furthermore, the potential productivity of the new machines could pose problems for the remaining conventional machine operators; for example, if many jobs were transferred to the new machines, conventional machine operators might not be able to make their bonus, and thus would suffer a loss of earnings. Besides these immediate and relatively 'defensive' questions, stewards also saw new technology as providing an opportunity for developing their strategy of obtaining involvement in decisions on investment. They felt that the union should be directly involved in the introduction of new technology rather than reacting to problems as they arose.

Independently of the AUEW, the white-collar union APEX also sought a new technology agreement, and it was this move which led management to consider their position on this question. Initially, senior managers expressed surprise at the AUEW's demand, since none of the company's other plants had a new technology agreement. Opposition soon gave way to seeking accommodation, however, and several months later a company policy document on new technology was drawn up. It was envisaged that this would apply to both the AUEW and APEX, and covered consultation over new equipment, redeployment and training, and redundancy. At a meeting to discuss this and other issues, the AUEW objected to having a single agreement, and gave notice that they intended to seek other improvements, such as job security and a reduction in working hours. Subsequently, management negotiated with the two unions separately.

Discussions between the AUEW and the company proceeded very slowly, and it was over three months before a joint meeting discussed the idea again. Shortly afterwards the chief engineer drew up a 'Proposed New Technology Agreement'. This provided for joint consultation over company plans involving new technology, including the selection and purchase of machines. Any subsequent changes in working practices were to be mutually agreed. Operation of the new machines was to be limited to skilled AUEW members, who were also to prepare programs. Other clauses covered training and remuneration. In addition, the draft document proposed that 'any data relating to the financial justification and manpower planning of new equipment will be made available to the Union'. While much of the document simply formalized established practices of consultation over major changes in working practices, this clause would have opened the way for more decisive intervention by the union in investment decisions.

The AUEW's first draft of an agreement replicated many of the chief engineer's proposals, with modifications in some clauses to strengthen the union's position. In addition, the AUEW sought a 39-hour week and the right for any operator who so wished to be trained on the new machines. To cover for the eventuality that conventional operators' bonuses might be adversely affected, they proposed that an agreed 'surplus' from CNC output be credited to ordinary machine operators' bonuses. The union also introduced a section on health and safety, which included the proposal that maintenance of the new equipment would be subject to mutual agreement. This draft was presented to management for negotiation at the first formal discussions held some six months after the chief engineer had put forward his proposals. Several areas of disagreement emerged.

The main issues of contention concerned job security, the protection of piecework bonuses, and union control over maintenance. Management rejected the attempt to link the introduction of a shorter working week to new technology, and proposed that a joint committee be set up to look into the problem of people displaced as a result of the new equipment, if and when that happened. They also suggested that a joint committee would look into bonus problems if it was discovered that conventional machine operators were suffering losses as a result of the new machines. On health and safety they were only willing to state that there would be consultation, and rejected the union's attempt to extend its formal controls. Other modifications were also sought, which would have had the general effect of making the agreement less precise. But these and other areas of disagreement quickly became irrelevant.

Up to this point negotiations had only involved the spares' division management. Soon after the discussions outlined in the preceding paragraph, the personnel manager sent a copy of the areas of agreement and disagreement to the personnel manager at the company's headquarters. This reflected the normal practice whereby the central personnel department co-ordinated divisional policies to ensure that no serious inconsistencies arose. However, on this occasion the central personnel department rewrote the whole of the proposed agreement and sent back a 'final position' for the spares management to present to the AUEW. This was a considerably weaker document from the union's point of view. In particular, it did not contain the clause covering disclosure of information about financial and manpower planning in connection with new machines.

Negotiations continued informally, but it was soon apparent that agreement could not be reached on the new document. A year after its initial response to the union demand for an agreement, management decided to end negotiations with both the AUEW and APEX and issue a new technology policy statement instead. This signalled an attempt to return to the policy outlined in its first document. The AUEW convenor threatened to call a strike if the policy statement was publicized. This threat probably carried some weight because a previous attempt to go over the stewards' heads by issuing a document to members had nearly resulted in a strike. The policy statement was not issued, and within a month negotiations had been resumed, based this time on the company's 'final position' draft. Discussions took several months and the resulting agreement was not formally signed until some time later—over twenty months after the union had first sought the agreement. It covered the same ground as the chief engineer's initial draft, with the addition of clauses covering health and safety, and a commitment by the company to discuss future problems of transfers and bonus losses arising from the new equipment. The union also managed to introduce a new clause providing for an annual review of 'progress on new technology'. Subsequently an addendum was agreed, without any significant opposition on the part of management, providing for the union to receive information about the number and types of jobs performed on the new machines. This addendum enabled the AUEW to monitor the movement of work from old to new machines fairly precisely. However, the massive transfer of work that had been anticipated did not occur, mainly because fewer machines were introduced than had originally been planned. Hence this monitoring was less significant than it might otherwise have been.

The union's aim of obtaining increased information flows had been thwarted early on in the negotiations, but this was compensated for—to some extent—by high-level contacts between the convenor and certain managers. These provided the former with much of the data he required, albeit within the limits of confidentiality. Some information received also illustrated the limited nature of the new technology agreement's provision for consultation. Through informal contacts the convenor learned that the company was considering plans to introduce new types of micro-electronic equipment about which he had not been officially informed. He felt prevented by the confidentiality of the information from pressing for the new technology agreement to be applied, but was confident that if the plans came to anything the union would be able to handle any problems at that stage.

Other issues covered by the new technology agreement,

such as displacement of workers and loss of bonus earnings, had not arisen by the time of our research. Nevertheless senior stewards felt that the agreement was useful if only because it provided them with the opportunity to quiz management about new machines at the annual review. This took place during wage negotiations, and the minutes of these meetings indicate that discussion ranged over performance problems, the implications of these for overall company plans, and future plans for CNC machines.

About half-way through the discussions on the new technology agreement, the union produced the 'AUEW Three Year Plan'. It explicitly linked the introduction of CNC machines to the improvement of conditions. This was a union reaction to the company's three-year plan, and received strong support from the membership in the discussions before the annual wage negotiations late in 1981. The union's plan looked, first, to achieving 'a 37-hour week without additional cost and increased production, through the introduction of CNC machines'. It also envisaged attaining extra holidays and holiday pay, consolidation of overtime pay, and revision of the PBR (payment by results) scheme. The plan was ratified by the members soon after the new technology agreement had been signed, and then placed before the management. The issues raised by the union were subsequently taken up during annual wage negotiations, and a joint working party was established to look at reform of the bonus system. This body had not reported by the conclusion of our research.

Work organization and the CNCs

As was stated above, the new technology agreement was of marginal significance as far as the planning and introduction of the new machines in the plant at the time of our study were concerned, since the negotiations had not been completed. We need, therefore, to look at the extent to which the union influenced management thinking on the selection of machines and the way work on them was to be organized.

The AUEW first learnt of the company's intention to introduce CNC machine tools during the 1979 wage negotiations. The union's immediate response was to argue that any benefits from the productivity of the new machines should be distributed equally across the whole workforce. They were opposed to any special treatment of CNC operators, as had been the case in another of the company's plants, since that tended to encourage sectionalism among members. It was later agreed that the CNC operators would be guaranteed their previous level of earnings. In effect this was a guarantee to pay them the earnings of the top-rated PBR workers, an arrangement which was reaffirmed in subsequent wage negotiations.

While, as will be discussed below, some managers were in favour of creating a separate CNC section, three factors indicate that management were quite happy to concur with the union's position that CNC operators should be treated no differently from other operators. First, certain key managers had past experience of problems with 'élitist CNC programmers' elsewhere. Second, the company could have rated the CNC jobs in order to apply the PBR system, either by work study or the development of the system of synthetic standards applied to conventional machining work. However, there was no work study department in the spares division, while to have set the planning department to adapt the synthetic system would have raised indirect labour costs. The latter was one thing the managing director in particular was keen to reduce, and so he would not have supported any move of this nature. An additional factor in trusting operators to work at satisfactory levels was the perspective, held by the managing director in particular, of harnessing operators' skills and interest rather than seeking to impose direct control over them. The AUEW, for its part, did not oppose placing the CNC operators on guaranteed earnings, since this fitted their long-term goal of doing away with the PBR system. Indeed they reckoned that, if the full complement of CNC machines planned had been introduced, more than half the operators would have been on a guaranteed day rate (since many other small departments had also shifted off the pure PBR system). Thus their objective of ending the system would have been substantially achieved regardless of what the results of any negotiations had been.

The union insisted on operator programming as a precondition of CNC machines being introduced. The management did not initially give any formal commitment to this, but to all intents and purposes did so in the new technology agreement. Indeed, as we show below, full operator programming of the CNC machines, particularly the machining centres, developed only over a period of a year or two. Before looking at this aspect of work organization, however, it is useful to examine the union's influence over the actual introduction of the new machines.

The AUEW had no direct input on the development of the company's plans for introducing CNC machine tools, although it was involved in the selection of the machines. The usual practice, when new machines were being looked at, was for the chief engineer and members of the production planning department to examine manufacturers' brochures and view selected machines. Once they had made their initial choice, they organized another visit, this time accompanied by stewards and operators. From management's view, the value of operators' opinions related to possible safety and operating problems. From the union viewpoint, however, the criteria used by operators and stewards in their assessments also included the potential for operator programming and the prospective benefits for the company generally. Nevertheless, it seems that, with one partial exception, worker and union views had played little role in the selection of machines. The stewards, for example, felt that some of the machines purchased were not powerful enough to machine certain types of steel, and as a result could not be used efficiently. However, this view led to no action on the part of management. Apart from this, the steward body was generally satisfied with the level of consultation and influence it had achieved, despite some problems of inadequate information. Machining centre operators said their requests for tooling had usually been acted upon, particularly in the first year or so. However, their opinions about a suitable shop-floor mini-computer had been ignored. Overall, it seems that the union's influence depended to a significant degree on the coincidence of their views with those of key managers—or the ability of the union to persuade them.

However, the AUEW played an important role in relation

to the company's CNC strategy by holding section meetings on the shop-floor to explain the plans to the workforce. These (without the workforce details) were initially presented to a meeting of the whole plant by the managing director at the annual communication meeting. He later met the shop stewards' committee to discuss the plans in more detail, and the stewards then organized meetings to inform members. As the convenor put it, the union sold the company's plans to the shop-floor.

The union's motive for doing this stemmed from the shop stewards' committee's strategic perceptions. The committee adopted the view that the best way to secure their members' interests, in terms of improved pay and conditions and job security, was to ensure that the company operated profitably. As the ex-convenor put it, trade unionists were unreasonable in restricting production for most of the year, and then seeking a slice of the profits. He felt that the union should recognize they had a responsibility for production and adopt policies accordingly. Some stewards had favoured a more confrontational approach, but had agreed that for the present they should seek their objectives by ensuring that production was carried out as efficiently as possible. As a result, when they saw management's projections of double or even treble the current level of output from the CNC machines, they felt that their introduction would provide an opportunity for achieving some of their other objectives; these were outlined in the 'AUEW Three Year Plan' discussed earlier.

When the first 'experimental' CNC machining centre was installed, it was set up on its own away from the main machining shops and close to the canteen. It was placed there with the intention of making workers familiar with such machines and their capabilities. For the first few weeks, parts made on it were displayed with labels showing the times taken to make them on the new and old machines. The former time was always much shorter.

The production planning department was initially responsible for this machine, even selecting an operator from the milling shop to work it (normally they had no direct role in shop-floor work organization). The milling shop supervisors were later asked to supervise and service the machine, but refused on the grounds of impracticality, since the machine was not near their section and because they were annoyed at the manner in which it was introduced. As a result, the first machine was supervised and serviced directly by the production planning department. It was not until the third machining centre was brought to the site that all three machines were moved into the milling shop and placed under the shop-floor supervisors' control. Subsequently all the new machines were placed directly in their appropriate conventional machine shops, although, as has already been indicated, the illusion of a 'CNC shop' was created by placing the machines on opposite sides of an open gangway.

The conflict between the production planners (under the chief engineer) and the shop-floor supervisors (under the factory manager)—both groups belonged to the same union, TASS—was less over the siting of the machines as such than the manner in which production planners had encroached upon the supervisors' sphere of control. The issue of the actual siting of the new machines, however, was a matter of contention between some managers, who wanted to have a CNC section separate from the conventional machine shops, and the AUEW, who were determined that the new machines should be integrated into the existing shops. The AUEW viewed the final outcome as an acceptable compromise, but, since it never exerted any overt pressure on the siting of the new machines, it seems that key management groups also moved to support the solution finally adopted.

The major question with CNC machines is, of course, the control of programming. All the managers interviewed said that they had favoured some operator programming from the outset, partly because there was no separate programming function in the factory, and partly because they recognized the need to utilize operators' traditional skills. However, there were also clear indications that the management as a whole had not started with such a straightforward perspective, and/ or had had different ideas about the degree to which operators should have sole programming responsibility. Up until mid-1981, management documents still treated the degree of operator programming as an open question, and indeed frequently referred to a division of programming between the operators and the production planning department. In one document, for example, it was proposed that the production planning department would be responsible for 'complex' computer-aided programming, while operators would perform all other programming. Members of the planning department favoured this division of labour. For its part, the AUEW had always sought full operator programming on the shop-floor. However, they had been able to achieve this only some three and a half years after the first CNC machine had been brought into the plant. The senior steward was unable to account for the delay, except in terms of disagreement among managers over operator programming. By early 1982 the AUEW had the company's commitment to operator programming enshrined in the new technology agreement, and the union confined itself to reminding management occasionally of their commitment on this issue. The move to full operator programming speeded up after the appointment of one particular manager to a senior position. The convenor knew from informal contacts that he had the support of senior managers, and at no stage did the AUEW consider exerting any pressure on the company on this issue. Hence, the union was prepared to wait for the disagreements within management to be sorted out, confident that they would achieve their goal. Such a policy also avoided a direct confrontation with the TASS membership in the production planning department. By 1983 full operator programming was achieved. It was symbolized by the installation of the mini-computers on the shop-floor early that year. At the time of our research, senior managers expressed themselves happy with full operator programming. It was seen to lead to high shop-floor morale, provided access to shop-floor skills, and eliminated the need to prove office-made tape programs. In short, experience had convinced management that operator programming was feasible for small-batch production, and so there was nothing to lose by moving in the direction the AUEW wanted.

Crucial to control over programming are, first, the acquisition of knowledge, and hence training, and second, access to the relevant equipment. Training to use the CNC machines never seems to have been a problem from the point of view of those regularly operating them. They all had several weeks' training, including attendance on manufacturer's courses. The main complaint made by the AUEW regarding training was that not enough back-up operators were trained initially, and that facilities for these workers to gain working experience were inadequate. Having failed to obtain any agreement for further training at factory level, the union had approached supervisors and, at the time of our case-study, had reached an informal agreement with the turning shop supervisor, which would allow back-up operators several hours' CNC work experience each month. No such agreement had been achieved in the milling shop, however, by the end of our research.

The extent of operators' programming knowledge varied considerably from one worker to another. While all the operators could and did program their own machines, one machining centre operator in particular had developed programming skills far superior to anyone else in the factory. Some other operators' knowledge also clearly extended beyond that required for day-to-day programming. The most skilled operator, who had been transferred to work the first CNC machine, had had additional programming training, and was also personally very interested in computing—to the extent of working on problems at home. Further training in programming was being given to other operators, partly by the skilled man, although it was not organized very systematically. Another operator had only had additional training through being sent on a manufacturer's course for a machine that the company, in the end, did not in fact buy. However, this proved more than adequate for him to start using a newly purchased machine before the manufacturer's representatives could come to the factory. When they did arrive, he was able to demonstrate that their machine had more capabilities than they were aware of.

Reference has been made earlier to the later machines having a simpler 'menu' type of programming system. Shop stewards discussed the implications of the later machines for operator skills when they were introduced. They felt that there was probably some loss of skill overall, especially compared with the older CNCs, but were equally convinced that such programming systems were necessary for spares production if the projected efficiency levels were ever to be attained. As we have seen with reference to the union's three-year plan, the AUEW considered that increased efficiency was necessary for the attainment of other objectives, such as improved wages and conditions.

Access to programming knowledge, though uneven and unsystematic, does not seem to have been a problem. There was unrestricted access to programming equipment on the machines. Not only were facilities for programming machine operations open to the operators from the start, but operating programs supplied by the manufacturers were also accessible to them. One operator, who did not have any additional programming experience or training, altered these operating programs within a few weeks of the new machine being installed. He explained that the machine's parameters had been set to be suitable for supervision by an unskilled operator, and he modified them to make better use of the machine.

Access to computing facilities off the machines was initially less easy for the shop-floor workers to obtain. The time-share computing equipment was initially installed in the production planning office, which was adjacent to, but quite separate from, the shop-floor. Mini-computers were also installed in this office, and remained there until early 1983. Up until this time the planning department made some machining programs. One member of the planning department began specializing in programming, and experiments with office preparation of computer tapes were conducted. Moreover, the location of additional computing equipment in the planning office meant that complex programs had to be made there. This task was sometimes performed by production planners alone, and sometimes by the more skilled of the shop-floor operators, who would have to go to the planning department. As was discussed above, in 1983 the agreement on full operator programming was put into effect.

Two other aspects of the introduction of CNC machines are also worthy of brief comment here. The first of these concerns CNC job time-sheets. These were to be filled in by each operator for every job and were introduced when the first machine was brought into the factory. At that stage it had been suggested by the chief engineer that jobs should be monitored more formally, but this was never done—probably because close monitoring would have required setting staff to do that work, thus increasing indirect costs. For their part, the operators could only explain this in terms of the company's preference for trusting them over close monitoring. (The only check on these records was made by the assistant supervisor, who merely ensured that the times were arithmetically consistent—no checks on their accuracy were made.)

The second point concerns subsequent proposed developments. When it became apparent that the CNC machines, particularly the machining centres, were not working as efficiently as had been predicted, senior managers tried to introduce significant changes in working practices. They proposed that the machines should be left running throughout lunch-breaks, supervised by such operators as normally took their lunch on the shop-floor. These men could then take time off later to compensate for this work. They also proposed that the machines should be allowed to run until their programs had been completed at the end of the day. The AUEW strongly opposed both of these moves, primarily on the grounds of safety, and also the cost of replacing tools that would inevitably get broken if the machines were effectively run unmanned. They were also opposed to any departure from the practice of one man per machine without considerable thought and negotiation. As a result of this opposition, no attempt was made to introduce these changes, at least by the time we left the plant.

In this section we have looked at the way in which the actual pattern of work organization on CNC machines developed. We have seen that this matter had not been clearly thought out prior to the introduction of the machines and, indeed, that it was several years after the introduction of the first CNC that the final pattern was established. It came about due in part to union demands but also to the inclinations of certain key managers. Nevertheless, a central issue—that of programming—was an issue of some significant disagreement, as has been noted in other studies (e.g. Wilkinson 1983; Brady 1984: 56). In effect, the new equipment presented management with two alternatives. The key task of programming could be used to reduce the skill of operators and foster the development of a more managerially controlled pattern of production: this might be seen as increasing the degree of bureaucratic control. On the other hand, a strengthening of a form of craft administration was possible both by having operators program and by exploiting the fact that, since CNC machining centres undertook a variety of tasks, there was less need for a complex system of progress chasing. The latter had always been ritualistic to some degree, and became even more so with the introduction of CNC equipment since the planners lacked the ability to make optimum use of the new machines. In part because of the skilled nature of the work done in the factory and the desire to maintain the craft ethic, and in part in a desire to reduce administrative costs, the latter course was chosen (similar findings have been made by Jones 1982; Sorge *et al.* 1983). Moreover, management was increasingly thinking of using transfer machines and the like which would further serve to reduce the complexity of the bureaucratic or administrative structure.

Generally speaking, the union had a significant degree of influence over matters relating to work organization, although it should also be noted that they often had significant management allies. For example, CNC operators were not treated as a separate group; they did program; they were able to win a number of commitments from management on the future use of new technology; and they were able to prevent moves towards an intensification of effort. While they had little influence over the general strategy of management, they did accept its underlying philosophy and sought to gain improvements in wages and conditions out of it, with some success. As in previous chapters, we again find that the pattern of union influence broadly replicated its more general role. The major point of difference was that CNC machines introduced a new principle, which potentially challenged the traditional craft way of working. In order to protect craft status, therefore, the union felt it necessary to ensure that operators programmed.

THE GENERAL PATTERN OF LABOUR REGULATION AND UNION INFLUENCE

In outlining the pattern of work organization we have already touched upon many broader aspects of labour regulation. We shall not repeat those points here. Rather we discuss questions of labour allocation and more general matters relating to pay and conditions.

First, we consider recruitment. The number of shop-floor workers in the engineering plant had grown very rapidly to around 350 by the mid-1970s, since when there had been no increase. Some of the first workers came from other company sites, while the rest had been recruited locally. The union had exercised no control over recruitment, either when the factory was first set up or later. The convenor was satisfied that recruits had to join the AUEW, and that, since the company required skilled labour, only fully qualified men would normally be recruited. There had been virtually no recruitment of machine operators in recent years, but in the past it usually took place through advertisements in the local press. Occasionally temporary jobs did become available in the plant, and the union sought to have advertisements for these placed on factory notice boards, so that operators could inform out-of-work friends. In one case, the notices were put up, but almost at once the company took on people who had just taken redundancy from another factory and who were not known to any of the shop-floor workers.

Transfers of people between shop-floor departments were guided by a long-standing, though informal, agreement. Under this, the union had agreed that anyone could be transferred to any department provided the transfers were voluntary, and that individuals could return to their original department if they did not like the new work. Individuals were guaranteed their average earnings, or, if on time-rated jobs, might even be upgraded. The choice of individuals for transfer from among the volunteers was, however, left entirely to the departmental supervisors, who assessed volunteers on the basis of their skill. In some departments the shop steward was informed of the details of transfer plans, but did not intervene in any other way.

Promotion of a machine operator to the position of a leading hand usually occurred only when a supervisor was absent. In principle, departmental stewards were informed of who was being made-up, but in practice this was rarely done formally. Supervisors chose whomever they considered suitable. The only constraint the union imposed was that stewards should not be leading hands—though in one case they had to accept this. Shop supervisors were recruited from among the machine operators: the union exercised no direct control over this at all. However, on all of these issues of labour allocation and promotion, it was clear to everyone concerned that the stewards would become involved if an operator so wished.

Individual machine operators had some control over their earnings through the working of the PBR system, as has already been discussed. Time-rated workers in the warehouse and inspection department lacked such direct control, and their basic rates were supplemented by a bonus dependent on the overall level of machine operators' output. The value of the elements of the PBR system, basic wage rates, holidays, and working hours were all negotiated at spares' division level. Before looking at these bargaining arrangements, we should note that overtime and certain departmental allowances were regulated by site-level agreements.

The overtime agreement laid down two types of overtime, both of which had to be agreed beforehand with the convenor. Regular weekday overtime was to be available to everyone on the shop-floor, regardless of the actual work required. The specific amount of this overtime was negotiated when necessary. To meet management's need for further overtime in particular departments, the union agreed to recognize a category of additional overtime to be allocated by managers on an individual basis. This sometimes led to ill-feeling between departments, and to overcome this the union on one occasion suggested that men should be transferred from the shops with little work to those where regular 'extra' overtime was being worked. Management did not agree, but did increase the amount of overtime available to everyone. The convenor accepted this as it put financial pressure on the company to move people into the understaffed departments, while satisfying people's concern to earn more overtime in the short run. The only other site agreement concerned special allowances in the assembly shop for testing pieces of equipment put together there. This had been negotiated several years ago, and at the time of our research nearly everyone in the department was being paid this allowance.

Pay and other conditions were negotiated annually at division level, together with representatives from a related site. On pay, the union had been able to increase in real terms both the basic rates and the earnings ceiling of the PBR workers over recent years, and in addition had reduced the differential between time-rated and PBR workers' pay. Earnings were highly competitive for the area. They had also raised the question of a shorter working week, their objective being to have the same hours as clerical workers (negotiations had not been completed by the end of our study). The union side had also argued for a complete revision of the payment system, replacing PBR with a time-rate for machine operators, on the grounds that the present system was inefficient. As noted above, after several years this issue was finally taken up, the company establishing a joint *ad hoc* committee to look into the question. Similar committees had also been set up in the past on rate-check problems. The convenor or his deputy sat on these bodies to represent the union side. The only issue discussed at company level was the working of the company pension scheme. An AUEW member sat on the pensions committee.

Both formally, through a works committee, and informally, through strong bargaining relationships, the union was involved in consultations on a wide range of issues: these included management plans for the future, market trends, levels of performance, and a variety of production-related problems. In addition, the company had in recent years experimented with a variety of methods of involving workers more fully. These experiments included the use of quality circles and briefing groups, along with other communications exercises. These appear to have had little impact, with the result that they tended simply to die out after a period of time.

In overall terms, the union achieved a significant degree of influence compared with many British workplace organizations. While there was a certain amount of consultation over strategic issues, the union played a role in many aspects of labour regulation. In addition, and in part due to the structure of union control, individual PBR workers were able to engage in a significant degree of bargaining both over work arrangements and their own pay (although the scope for individual bargaining over pay was less for CNC operators and others not working on piecework). This pattern of labour regulation—characterized by a dependence upon craft skills and supported primarily through financial incentives and a fostering of the craft ethos—was also dependent upon the nature of the organization of the AUEW within the plant.

UNION SOPHISTICATION

The AUEW was the only union representing manual workers in the plant, and had a sophisticated organization. The basis of this was 100 per cent membership, with a *de facto* closed shop in operation from the initial establishment of the plant in the early 1970s. Agreement had been reached between management and local AUEW officials at this time that the union would represent all shop-floor workers on the site. As a result, the AUEW membership had grown with the expansion of the plant, reaching a peak of about 350 members in the mid-1970s. Since that time both employment and membership had remained stable. Nearly two-thirds of the members were setter-operators. The remainder consisted of labourers, warehousemen, inspectors, paintshop workers, and maintenance workers, including electricians.

The second aspect of union organization of importance in understanding the role which the union was able to play was the distribution of shop stewards. The company had agreed from the outset that there could be stewards, and early agreements outlined the facilities to which they would be entitled. Stewards were elected annually, and initially there had been a relatively high turnover since they were often promoted to supervisory positions. From about the mid-1970s, when the labour force ceased to grow rapidly, a more stable structure emerged under the leadership of a handful of activists. Since that time, turnover among stewards had been low, and was typically due either to promotion or resignations for personal reasons. There had been only one case of competition in the elections for stewards in the last few years. For several years before the research, the fitting and assembly shops formed one shop steward constituency, and the steward came from the assembly shop. Some of the fitters gradually began to feel that the steward was out of touch with their needs. The fitters, unlike the assembly shop men, were on individual bonuses at the time. These were becoming harder to obtain as a result of a range of production changes. Then the two shops were separated physically during reorganization of the shop-floor, and this made contact with the steward harder (though there were no barriers to the stewards moving between departments). As a result, when the time came to elect a new steward the fitters put up a candidate, who was elected in place of the previous steward.

The ratio of stewards to members agreed with the company was 1 : 30, which made for a total of eleven stewards. The convenor ensured as a matter of principle that shop steward constituencies broadly corresponded with the main production units, except in the milling shop, where there were two stewards because of the size of the shop. None of the CNC operators was covered by any special steward because of the principle that they belonged to the major machine shops. Until recently the company had stuck very rigidly to the ratio of stewards to members initially agreed with the union. For example, in 1980 the first convenor resigned, and the current incumbent, who had been a steward in the turning shop, was elected. The stewards felt that it was inappropriate for the convenor to be the steward for a major shop at the same time, and therefore asked for the number of stewards to be increased to twelve. The company refused, however, and the union's problem was eventually overcome by removing the general labourers' steward constituency and then electing another turning shop steward. As a result, the majority of the stewards were machine operators, as was the convenor. However, a new constituency had been created more recently for the assembly shop, which had been growing in size with the transfer of a number of men from other departments.

A third important dimension of workplace union organization concerns the integration and co-ordination of members and stewards. This was achieved in a number of ways at the plant. A shop stewards' committee met regularly every month, chaired by the convenor. He was also responsible for drawing up the agenda, although other stewards could raise such items as they wished. Regular items included reports and discussion of works committee meetings, safety reports, major issues raised with management since the previous meeting, and a report of his activities by the convenor. A somewhat unusual feature of the committee was the principle that each steward should be given particular responsibilities beyond representing his own constituency. Hence, in addition to a convenor, deputy convenor, secretary, and treasurer, other posts included an apprentice liaison steward, a pensions representative, safety stewards, and welfare and canteen representatives.

The convenor was a key figure in the co-ordination of the union within the plant. Under an agreement with the company, he worked full-time on union business. In addition to chairing the shop stewards' committee, he sat on the works committee along with his deputy and another steward chosen by rotation. He enjoyed strong bargaining relationships with a number of key managers. Stewards constantly came to him to raise problems, discuss issues, and seek his help and support. He also had frequent contact with members, preferring, for example, to have an office on the shop-floor rather than somewhere which would have made member contact more difficult.

The convenor, supported by his deputy and the secretary, formed a group of senior stewards, and were seen as an influential group of opinion leaders, a 'quasi-elite' (Batstone *et al.* 1977), by most people in the plant. Often in consultation with the other members of this trio, the convenor—who had previously been deputy convenor—played a key role in initiating union policy and action, whether over annual wage claims or other issues, such as the union's three-year plan and the new technology agreement.

Systems of integration were not confined to the steward body, but also spread out to close links between the stewards and the membership. They had, of course, a good deal of contact in the course of the working day. In addition, more formal meetings were held. Plant-level meetings involving all manual workers on the site could be held monthly in working time. They were usually held three or four times a year, mainly to discuss wage negotiations. The progress of wage talks was reported at these meetings, which also agreed any significant changes to the claim and ratified the outcome. Plant meetings had also been called in recent years to seek support for a one-day strike in support of health service workers, and to present a report on the state of talks over changes in the length of the working week. Plant meetings were convened by the shop stewards' committee, and by convention were addressed only by the convenor from the shop steward body.

Meetings of workers in individual shops could also be held, being convened by the shop steward responsible. The incidence of such meetings varied from one shop to another. In the grinding and fitting shops, for example, they were held only occasionally, in the first case because there were very few workers, but in the second because the steward preferred to go round and talk to workers individually. He held shop meetings only when he felt it would be a useful way of making a point to shop supervisors. In the turning shop, meetings were held regularly (about once a month) and less frequently in the milling shop.

Shop meetings could be concerned with a wide range of issues. Wage claims were drawn up in the first instance by the convenor, and circulated to stewards, who usually took them to shop-floor meetings for discussion. The results were then taken to the shop stewards' committee, where the shape of the final claim was finalized. Shop meetings were also called to discuss production information from the company. When the company announced its three-year plan to introduce CNC machine tools, the details were taken to the shop-floor by stewards, who held shop meetings to outline what they saw as the advantages of using new technology. The AUEW's three-year plan was drawn up on the basis of ideas aired at a series of shop-floor meetings following the company's announcement of their plan to introduce a large number of CNC machines.

Occasionally other meetings of groups of workers were held, such as one involving the turning shop CNC operators (and back-up operators) and departmental supervisors. This was called by the convenor to discuss the problem of access for back-up operators to the machines. It was agreed that further meetings would be held to iron out any other specific problems (although by the end of our research none had been held). CNC machining centre operators in the milling shop, however, organized their own meetings without the assistance or co-operation of the shop steward or the convenor. At these meetings they discussed issues concerning tooling and work arrangements on the new machines, and had drawn up a document listing these problems which they presented to the supervisor. There was no evidence that managers encouraged such behaviour on the part of the machining centre operators.

The final method of member co-ordination and involvement was through a regular series of leaflets produced by the convenor which covered issues under negotiation, the situation regarding disputes (e.g. concerning overtime), and production prospects. A regular item in these newsletters was information about orders received, the timetable for the introduction of new CNC machines, and the general level of productivity and output in the spares division. Notices were also sometimes used to gauge members' feelings about an issue. It was more usual, however, for such communications to take place directly through mass or shop meetings, or through the shop stewards.

The preceding account indicates a high degree of intra-union sophistication. Union membership was total, and shop stewards were relatively active at both shop-floor and plant level. There were a number of senior shop stewards and a full-time convenor. Shop-floor and plant-level union activities were relatively well co-ordinated and, in addition, there were clearly strong links between the stewards and the membership. We need now to go on to consider the degree of external integration—that is, links with the larger union and shop stewards in other parts of the company. We shall also briefly mention relations with the non-manual unions in the plant.

The AUEW in the plant co-operated with the AUEW at another spares facility belonging to the company over wages and other conditions common to the two plants. Such co-operation took place through occasional joint meetings before negotiations and considerable informal communication throughout the year. However, there was no other form of regular joint union organization involving AUEW groups within the company. Up to the late 1970s a joint AUEW combined committee had met quite frequently. This committee was not recognized by the company, and stewards had attended meetings in their own time, financed through collections. The committee enabled stewards to co-ordinate policy over, for example, wage claims. This organization broke up when the spares group of plants was offered a pay rise, which was accepted by the stewards there but not subsequently offered to other workers. Since then only one or two combined meetings had taken place. Other than the fact that some stewards attended their local branches, links with the larger union were weak. The local full-time official, for example, was rarely, if ever, involved in negotiations with management in the plant, nor was there much discussion and information exchange between the convenor and the local officials. However, they had played a more important role in the early days of the plant, before the steward organization became more firmly established.

Up to 1976 there had been site bargaining with parity between the sites, established through senior stewards co-ordinating claims. Then for one year there was company-level bargaining for the first and only time. This had ceased with management offering a favourable wage increase only to the spares division workforce. Steward co-ordination across the divisions had taken place since 1977 only in response to problems of plant closures and redundancies, over which industrial action had been both threatened and employed.

On the case-study site there were two other unions beside the AUEW—TASS and APEX. The former organized the supervisors, assistant supervisors, and members of the production planning department. Clerical and other white-collar workers belonged to APEX. None of the three unions had any formal or informal links with each other. It can be seen, then, that the degree of external integration was rather limited—although sufficient for solidaristic action to be employed—while union co-ordination across the manual–

non-manual divide was non-existent. However, the high degree of sophistication of the AUEW was an important factor explaining the considerable range of union influence.

In this chapter we have focused upon the introduction of CNC machines, a move which was stimulated by concerns with efficiency in the context of the need to improve the spares service to customers, and technical 'knock-on' effects from the use of such machinery in the production division. The details of the way in which the CNC machines were to be used took some time to develop even after their installation, particularly as far as the key question of programming was concerned. Finally, operator programming was established, in part due to union pressure but also due to the growing commitment within management to stress a form of craft administration, as against a more bureaucratic and hierarchical form of control. The exact implications of CNC machines for work experience depended upon the precise nature of the machines themselves. Traditional craft skills were still necessary, although they were typically used less frequently in a conventional manner. In order to program, however, operators had to conceptualize their traditional activities, in addition to learning how to program and, in some cases, undertaking a wider range of tasks. In addition, the payment system ceased to be the form of control which it was on conventional machines.

The union had, relatively speaking, a considerable degree of influence over matters concerning the use of CNC machines, as distinct from the decision to employ them—over which they were merely consulted. In addition to the question of programming, they played some role over the location of the machines, operator training, and the treatment of CNC operators. They also prevented moves towards an intensification of work. In addition, they used the introduction of new technology to put forward a wider range of demands, and insisted that the gains from the use of the new equipment should be spread across all workers. Such an approach on the part of the union reflected its more general pattern of influence, and was attributable to its sophisticated organization. This fostered unity and a sense of common identity among both members and stewards.

5

RATIONALIZATION AND WORKER AMBIVALENCE: NEW TECHNOLOGY IN AN INSURANCE COMPANY

IN this chapter we discuss the changes in a large and long-established company associated with the introduction of on-line processing of personal lines insurance, i.e. individual householder and private motor policies (as distinct from commercial premises, motor fleets, commercial vehicles, etc.). This change had started in the mid-1970s, and by the time of our research in 1983 had largely been completed, although some further developments were still occurring. Partly through the take-over of two other companies, total employment in the company had increased considerably during the 1970s but had then fallen by about 3 per cent. Some of this job loss was the direct result of the new system of on-line processing. The greatest job loss, however, was due to a variety of attempts by the company to cut costs and become more competitive. These, along with the transfer of membership from a staff association to form a company-specific division of BIFU, are crucial factors in understanding the history and effects of the move to on-line processing.

In the first section of this chapter we look at how work organization changed with the introduction of new technology and the considerations which led management to introduce on-line processing. The second section discusses the way in which management approached the labour aspects of this change, and goes on to assess the approach and influence of the union. In contrast to the other case-studies reported here, we find a significant expansion in the role of the union, although it still remained limited, particularly as far as the details of work organization were concerned. This, as the third section indicates, reflected the more general part which the union played in the company. The fourth section discusses the factors associated with both the expansion and limitation of the role of the union. Crucial to the former was not only the way in which rationalization raised new issues, but also the organizational developments associated with joining BIFU. Central to the latter were weaknesses of organization at local level, which constrained the role which a sophisticated company-level organization could play. These were in turn attributable to membership ambivalence concerning the balance between collectivism and individualism.

Before looking at changes in work organization associated with on-line processing, a number of points should be made concerning the nature of this case-study. Our other studies concentrated upon changes within individual establishments. In contrast, our insurance case-study took place primarily at company level. This was so for two reasons: first, the new technology was part of a company-wide system, linking area offices to the central computing centre. In other words, particularly with the on-line system, the local office was part of a national network. Overall, about two-thirds of the staff of the company were employed in the area offices, the remainder being based at company headquarters and in a central computing unit. Within the area offices, different sections dealt with different types of insurance. The focus of our concern—personal lines—was one such section, which in all employed about 250 people at area level. The total size of area offices varied between 40 and 100 staff, who worked under an area manager supported by a number of assistant managers.

The second reason for focusing upon the company as a whole is that bargaining was highly centralized. Hence, in order to look at the role of the union in the process of technical change, we had to concentrate upon the company level, although we also investigated bargaining at lower levels. (As well as observation and informal interviews, we undertook a survey of half (48) the union representatives in the company, in order to further our understanding of the pattern of industrial relations at local level.) The company focus of our work also means that the variations in organization between different area offices become particularly striking: work organization had varied significantly before the introduction of on-line processing, although the company was seeking to reduce these variations in order to develop a common pattern of 'best practice'. At the time when on-line processing was introduced, there still remained considerable differences between offices. As a result, the following discussion is in fairly general terms, stressing the common features of work organization.

CHANGES IN WORK ORGANIZATION

In order to understand the pattern of work organization, it is necessary to have some basic idea of the process involved in insurance. The starting-point is clearly that customers should wish to insure themselves or their property with the company. A great deal of the company's business was obtained through insurance brokers, and therefore close links with them were of central importance. Brokers typically deal with a variety of insurance companies, and it is in the interests of any individual company to ensure that brokers who handle a great deal of insurance pass business to it rather than to its competitors. A number of factors are important in this relationship, as well as the commission paid. Brokers need to be able to rely upon a speedy and helpful service from an insurance company, and wish to place their high- as well as their low-risk business. In addition, favourable terms on one minor insurance policy may sometimes be a means of securing much more valuable business for a particular company. These facts impose certain constraints upon the routinization of insurance work.

The key processes within an insurance company for our purposes are underwriting and producing the documentation relating to policies, both for the client and for the company. Classically, an underwriter assesses the risk attached to any proposal, and on that basis determines the appropriate premium, the price the company should charge to insure the person, piece of property, or whatever. The underwriter therefore has to keep an eye open for any possible complications in a proposal, and may on occasion deem it advisable to place certain restrictions on the cover given. The discretion which underwriters have varies both between different types of business and over time. In the company we investigated there had been a steady reduction since the 1950s in the amount of discretion underwriters enjoyed. Guide-lines and rules concerning the evaluation of risk and hence the appropriate premium had been developed. This was particularly true of personal lines insurance, where a very large number of policies were handled (so that risks were more widely spread) and where the premiums were typically small. It was not economic to investigate risk in great detail. Hence it was necessary to have standard policies with relatively easily applied rules, which were supposed to highlight the key factors associated with the chances of claims being made upon the company. The pressure for such simplification—with its associated economies—grew as competition from foreign companies increased. In contrast, commercial insurance involved much larger premiums, potentially much larger claims, and fewer policies. In this instance work could less easily be routinized.

Even before the introduction of on-line processing, then, there were fairly detailed guide-lines concerning the underwriting of personal insurance. In the case of car insurance, for example, the underwriter relied upon a series of manuals. These provided details of premiums for particular types of car, which were then adjusted according to the area in which the car was used, the person holding the policy, and a number of other factors. In other words, the key underwriting decisions were made in the construction of the manuals, a task undertaken at head office. Considerations of the kind outlined earlier concerning brokers meant that on occasion the underwriter might decide upon a premium other than that indicated by the manuals, although within parameters laid down centrally. The underwriter, then, following the guide-lines, and occasionally modifying them, had to calculate the premium which the company would charge to insure a particular car. Two processes were then necessary. The first of these was to put all the relevant details of the new policy into a standard format for the company's own records. Under the old system, which had been introduced in the late 1960s, this meant that forms had to be completed which were then sent to the central processing centre to be put on the computer. In addition, the documentation relating to the policy had to be filed at the office to permit future reference. Second, cover notes and later the full policies had to be written up and sent to the client. In short, the process of writing a new policy under the batch processing system involved a series of arithmetical calculations guided by a manual, and then a series of essentially clerical tasks.

The personal insurance section at an area office was also involved in other tasks. Important among these were changes to policies. For example, several months into an annual policy a client might change his car, so that the insurance policy required amendment. The underwriter therefore had to calculate the annual premium for the new car, and then adjust it for the period of the current policy which was still outstanding, thereby obtaining a figure for any additional payment required from the customer. Similarly, policies had to be renewed each year. The paperwork involved in these two activities was essentially the same as in issuing a new policy. However, a personal insurance section at area level would also be involved in answering queries from clients, both directly and through brokers. These might come by post or telephone. Very often they involved finding the file relating to a particular policy (frequently a frustrating task) in order to answer detailed points. In other words, the process outlined in earlier paragraphs was subject to continual disruption, since telephone queries had to be dealt with.

In addition to looking at the functional tasks involved in handling insurance at area level, it is also necessary to consider how the performance of these tasks was organized between different members of staff. A major constraint upon any rigid form of organization was that the pattern of work was not totally predictable. Assuming there was no backlog of work, the tasks to be undertaken on any one day were determined by what came that morning in the mail, and what telephone calls were made during the day. From experience, of course, there was a general idea of the pattern of work, and peaks could be handled by allowing a backlog to develop. However, the first task each morning was the sorting of mail into different categories and its allocation to different members of staff. Under the old system, one major division was between different types of personal insurance; that is, certain underwriters and clerks specialized in, say, household insurance while others dealt with motor insurance. The exact categorizations of mail, and hence principles of task allocation, had varied between different offices. However, a primary aim was to differentiate quickly between matters of varying complexity. More routine work could be handled by less experienced and more junior staff, while matters requiring greater knowledge and skill—such as those involving greater underwriting discretion—would be allocated to more senior staff.

Under the old system, two sections were involved in personal lines insurance work at area level. The first of these was the personal lines section which was essentially concerned with the underwriting process. The detailed organization of the section varied between areas, reflecting the scale and nature of the insurance work undertaken. But the basic structure was as follows: the section was headed by a superintendent, possibly aided by an assistant. They supervised the work of the less experienced staff, allocated tasks, and checked a proportion of the work done by other staff. Below them were formally qualified and experienced underwriters, specializing in particular types of personal insurance. These were aided by underwriting clerks, who undertook more routine aspects of the underwriting process and in many respects were trainees. In addition, there was an underwriting services section largely concerned with the more clerical aspects of insurance. Within this department were underwriting services clerks, aided by clerical assistants. They were formally required to undertake the essentially clerical aspects of the process, but they might also assist in the underwriting process. The extent to which they did the latter varied considerably between areas, although many underwriting services clerks had some knowledge and experience of underwriting. It should also be noted that the batch processing system involved a good deal of highly routine work in the transference of data from policy forms on to the computer at the central processing unit.

Management reactions to market forces: rationalization and new technology

In the mid-1970s the insurance market became increasingly

competitive, and the commercial sector—traditionally the focus of the company—was becoming less attractive compared with the personal sector. Competition grew with the entry into the British market of foreign companies, which, using different underwriting and organizational techniques, were undercutting the major British companies. At the same time, high inflation meant that return on investments became a significant source of revenue for insurance companies. This additional income could be used to cover growing underwriting losses, which were in part associated with inflation. Increases in premiums could therefore be limited and hence competitiveness maintained. Competition made commercial insurance less profitable. However, when the recession hit, the commercial market tended to decline, in some areas by as much as half, as plants shut down and labour forces were reduced. This made the personal insurance market relatively more attractive, as did also the greater ease with which premiums could be increased in line with inflation.

A further, chance factor made our company especially aware of the potential of the personal insurance market. One consequence of the developments we have just outlined was that a number of small insurance companies had been set up: these often subsequently collapsed. In order to maintain market confidence, the large companies made an arrangement whereby they would take it in turns to buy these ailing companies, gradually running them down and incorporating their business. Our company was obliged to do this in the mid-1970s. It took over a company which was in financial difficulties, although was basically profitable. It had acquired a significant share in one part of the personal insurance market through using what at the time were advanced computerized methods to pare administrative costs and thereby premiums. This served to make the company studied even more aware of the viability of such a strategy.

These factors, then, led the company to rethink its approach to the personal insurance market. It developed two interrelated strategies. The first of these was to rationalize policies and make them more customer-orientated. By this time, in the mid-1970s, the company offered a wide range of overlapping policies for personal insurance, totalling about fifty in some areas of the market. These had been introduced over a long period of time and in an *ad hoc* manner in response to specific market opportunities, with little attention paid to the overall balance or to the profusion of types of policies which resulted. Furthermore, the way the policies were presented to potential customers reflected the intricacies of the insurance world, paying little attention to questions of comprehensibility and attractiveness to the layman. The first element of the company's strategy, therefore, was to reduce the plethora of policies to a single, basic set, and to ensure that they were written in 'plain English', thereby increasing their marketing appeal.

The simplification of policies, along with a further tightening up of underwriting rules, facilitated the second strand of the company's strategy—the reduction of administrative costs. The greater routinization of underwriting associated with these moves made on-line processing a more attractive proposition as a means of further reducing the costs of administration. (These strategies were common in the insurance industry at this time: see e.g. Barras and Swan 1983; Rajan 1984.)

The old method of dealing with personal policies had led to what one manager described as a 'paper explosion'. Tasks had become minutely subdivided, and the blossoming documentation involved with each policy was handled several times. This was not only costly and cumbersome but, it was found, also increased the probability of errors. On-line processing was seen as a means of simplifying procedures, speeding up handling, and thereby cutting costs. Under this system, the underwriters at the area offices, which were valued by the company due to their local networks of contacts, would use terminals to input data directly into the central computer. In addition, the system was designed to print automatically much of the documentation required in issuing policies and billing customers. In this way much of the clerical work of the underwriting services sections would be eliminated, as would that involved in inputting data at the central computing unit (cf. MSC 1985: 75).

The proposals to introduce on-line processing were developed by the systems development specialists at the company's computing centre along with the manager in charge of personal insurance, who himself came from a computing background. However, it took over five years to introduce the scale of on-line processing which existed at the time of our research. First, this key group of initiators had to convince a cautious senior management of the viability of their proposals. Second, the central specialists had to learn a great deal about the day-to-day tasks and problems involved in dealing with personal insurance, in order to develop an adequate system. Detailed studies were therefore made, and in addition selected 'users'—those who were to work with the on-line system at area offices—were involved in detailed discussions; one user was seconded full-time to the project team. The complexity of the process and the need to ensure that it was fool-proof led to the decision to introduce on-line processing incrementally, taking different types of policy and different tasks (such as new policies, amendments, etc.) one at a time. Even so, implementation took a good deal longer than initially envisaged. Users and pilots at each stage highlighted a considerable number of unforeseen and difficult problems. Labour considerations also led to some delay; these are discussed below.

Before looking at the pattern of work organization with on-line processing, it is also important to note that rationalization of policies, routinization of underwriting, and on-line processing were not the only means by which management sought to reduce administrative costs. In addition, detailed studies were made of other aspects of work organization, and in the annual budgeting exercises managers were often told to reduce their costs by substantial amounts. It was this latter approach which had the most dramatic effect on employment and effort levels.

On-line processing and work organization

By the time of our research, on-line processing had been associated with a number of changes in work organization in personal insurance sections at area level. If we focus first upon the more 'mechanical' aspects of task performance, six key changes are worthy of note. First, underwriters input data directly into the computer through terminals located on their desks within the area office. This eliminated the need—in most cases—to complete the lengthy forms which had previously been sent to the central computer unit for inputting. Second, with the on-line system the underwriter keyed in the details in an interactive mode with the computer. That is, a series of questions were displayed on the screen, and the underwriter had to fill in the relevant answers. The order of inputting was determined by the program rather than the underwriter. The system was designed to be user-friendly: that is, it was highly simplified, with little knowledge of complex codes being required and few questions being displayed on the screen at any one time. Third, the computer now undertook the calculations required to determine the premium. In other words, the underwriter was required to undertake fewer arithmetical tasks. In one sense this can be seen as a reduction in the intellectual demands of the job, but it was a change to which few underwriters were opposed. The computer calculated the premiums according to the guidelines; however, it was possible for the underwriter to override the computer so that he could give discounts etc. Fourth, a great deal of the documentation which had to be sent to clients was no longer produced by hand at the area office. The central computer automatically printed this out, and the documents were then sent to the area office for checking and sending on to the client. Fifth, the underwriter was able to use the on-line system to call up many details of policies. As a consequence, he did not have to make such frequent—and often frustrating—resort to the files when answering queries. Sixth, we should note that there was no longer any need for staff at the central computing unit to input data. (The following discussion, however, will consider only work organization at area offices.) In summary, the key feature of the on-line system was that it reduced the amount of paperwork. It also reduced the need for underwriters to do arithmetical calculations and to resort to the files. As a result it took less time to 'write' a policy.

However, there were also a number of broader changes in the way in which personal insurance was handled. It has been shown that before the introduction of on-line processing, underwriters specialized in particular types of insurance. Now they were required to deal with the full range of policies.

Hence the knowledge requirements of the job were increased. However, this increase was less significant than might at first appear to be the case, since the range of policies had been dramatically reduced and the underwriting rules had been tightened. Underwriters also needed to understand how to operate the new system: this, however, required relatively little additional knowledge, since the system was user-friendly. Hence, training to operate the system often took only a couple of days. Moreover, the speed of inputting was limited by the scant familiarity of most underwriters with the keyboard.

A further significant change was the elimination of the underwriting services section, with the staff being transferred to the personal insurance section, where they worked as underwriters. This was done since the on-line system dramatically reduced the amount of form-filling and documentation which had to be undertaken manually at area offices. At the same time, the new policies led to a substantial increase in the number of policies which were handled by the company, so that there was a need for additional underwriters, despite increased productivity.

While the terminal became central to the process of handling personal lines insurance under the new system, it would be wrong to suggest that it dominated the work of the underwriters. Indeed, given their cost, the company had not, at the time of the research, provided a terminal for each underwriter. Hence, staff had to organize their work in such a way as to be able to take turns at inputting their work or using it to answer queries. More generally, the documents associated with new policies and amendments still had to be thoroughly checked before inputting, and this took a substantial proportion of the total time involved in issuing a policy. Certain types of insurance had not been incorporated into the new system, and in these cases the traditional forms had to be completed in the same manner as before. Furthermore, while the files held in the office figured less centrally than they had in the past, they were still important. The on-line system did not incorporate all the details which might be necessary to deal with queries, and original documents still had to be filed for future reference. In addition, the need to deal with customers' queries remained. While the terminal was used to call up information, some underwriters still spent a good deal of time talking to clients and brokers on the telephone.

Effort levels also increased with the on-line system, particularly at one point a few years before our research. However, this should not be attributed directly to the on-line system, but rather to the cuts in manning relative to workload which came about through the annual budgeting exercise. In some cases it was realized that workloads had become excessively heavy, and accordingly additional staff were recruited. At the time of our research, despite some increase in effort levels, the atmosphere of offices was still relatively relaxed and easygoing. Very few people worked full-time with the terminals; the average daily number of policies input per underwriter was less than twenty. Finally, the exact pattern of organization in area personal lines departments continued to vary after the introduction of the on-line system.

Two other changes occurred with the move to on-line processing. The first of these was a general upgrading of staff. Hence, the number of area-level jobs in the highest three grades more than doubled, while the number in the lowest grade was almost halved. (However, the basic structure of grades and of supervision remained the same.) The second change was the elimination of a number of jobs. It is, however, extremely difficult to provide an exact picture of the employment effects of on-line processing as such, as distinct from the simplification of policies and the cuts associated with the annual budgeting exercise. It is certainly true that the number of policies handled per underwriter increased substantially. But this was to a large extent counterbalanced by the increase in the total number of policies. However, it is possible to provide a rather rough estimate of employment effects. The key phase of introducing the on-line system was associated with a reduction in staffing of less than 10 per cent (these being primarily low-graded jobs). These reductions were made through 'natural wastage'. On the other hand, the formal establishment levels in personal lines insurance at the time of our research were less than two-thirds of the number employed six years earlier, that is, before the move to on-line processing had been started. Not all of this reduction can be attributed to on-line processing as such. Hence, we can suggest that job loss directly attributable to on-line processing was somewhere between one-tenth and one-third. However, it should also be noted that the higher figure is a notional one, referring to standard staffing levels. Actual staffing had in the past typically been substantially below these levels. The reductions reflected a closer alignment of standards to reality (although the failure to staff up to establishment levels may in part have been due to the realization of forthcoming reductions in labour requirements).

There were, however, further job losses during our research as a result of a subsequent development. This was the establishment of a separate central processing unit for routine amendments of company policies in its subsidiary. The latter had adopted a very different strategy towards on-line processing, in large part reflecting the very specialized nature of its insurance business. Instead of locating on-line processing at local level, the subsidiary had centralized it. The nature of inputting work here was very different from that in the main company. It required negligible underwriting skills but a much greater sophistication in inputting data—the system was not 'user-friendly'. Moreover, the staff involved in the work were not dealing directly with customers or brokers. The work was therefore of a much more routine nature, and the input rate was far higher than in the main company— more than ten times as high per day per head. Management therefore decided that areas should send simple, routine amendments of policies to this central processing unit. These required minimal underwriting skills, and were therefore seen as particularly suitable for the skills and system developed in the subsidiary. This was expected to lead to a further loss of about thirty-five jobs. However, it also meant that the remaining staff within the main company undertook fewer routine tasks.

The move to on-line processing, then, led to some job loss and improved grading and pay, and also had a considerable impact upon the nature of work. In particular, the amount of paperwork and filing undertaken at area level was reduced, although by no means entirely removed. The rationalization of personal insurance associated with the move to the system accelerated the long-term trend towards less writing discretion, while underwriters had less freedom over the exact order in which they input data. Underwriters had less frequent need to use their own arithmetical skills, and less frequent resort to the files; but for many of the staff these were not seen as a disadvantage. For those who had previously worked in the underwriting services section the effects were somewhat different. Their work became considerably less routine as they became more fully involved in the actual underwriting process (for broadly comparable findings, see Storey 1986; Barras and Swann 1983: 47; Rajan 1984: 116–18). We need now to consider more fully how these changes in work organization and conditions were decided upon. In doing so we also look at the role which the union played in the move to on-line processing.

DETERMINING THE LABOUR ASPECTS OF ON-LINE PROCESSING

The basic decisions to reduce the range of policies offered by the company, to tighten up underwriting rules, and to move to on-line processing had important labour implications. To the extent that these strategies were aimed at reducing administrative costs, they were clearly likely to change work organization and manning requirements, since labour was such a high proportion of those costs. Indeed, the initial aim had been to double the business and halve the staff. While there was a substantial growth in business as we have seen, it was not on the scale initially expected and, at the same time, job reductions were significantly less than those initially planned. We turn to the question of staffing levels below.

An important decision at the design stage was that the system should be user-friendly, that is, it should not require any significant skill to input. This decision was based simply on the fact that underwriting skills were highly valued and there was a desire to minimize retraining, as well as to pre-empt any possible opposition to the new system. Management therefore did not want the underwriters to have to learn complex codes and to deal with a screen packed with data.

It was increasingly realized that the success of the new approach to personal lines insurance depended, not only upon the nature of the policies offered and the sophistication of the hardware and software, but also upon the more general organization of the offices and the staff involved. Management's concern for office organization was reflected in a series of detailed studies of it and a steady stream of proposals to improve office practice. At the same time, local managers were encouraged to find their own economies by the imposition of tighter limits on their costs and expenditure. This, we have noted, was the main factor leading to job loss. The quest for greater efficiency was only partly related to the introduction of on-line processing, and continued even after it had been introduced.

Concern over the staff was reflected in proposals to upgrade their jobs. Traditionally, personal insurance had been treated as the 'poor cousin' of the other insurance areas. This had meant that it was used as a training ground for staff, who then moved off to higher-status work. Consequently, management believed that 'problem staff' tended to be concentrated in personal lines insurance, blocking such career routes as existed there. This situation developed into a vicious circle, in which there were ever greater incentives for able staff to seek transfers to other departments. These problems were seen as potentially endangering the operation of the new on-line system. Upgrading was seen as a means of overcoming this risk, by improving morale and making this area of work more attractive. A second set of considerations, leading to the same upgrading of staff, concerned the need to avoid any opposition on the part of the staff to the new system. The logic for upgrading, from this perspective, was that it was a 'sweetener' to win staff co-operation in the key phases of the change. The union believed that this was the primary rationale underlying the upgrading process.

However, in their submissions to senior management the project team stressed a third theme. They argued that staff would be required to have a wider range of knowledge under the new system. When the change was completed, it was claimed, not only would underwriters have to be able to handle all types of personal lines insurance—household as well as motor insurance, for example—but the logic of the system also meant that an underwriter would deal with all aspects of the policy. The detailed division of labour which had been built up under the batch processing system was seen as incompatible with the on-line system (although there was a growing realization within the company that, even with the older system, a detailed division of labour meant that handling was slower and the error rate higher). Unless underwriters were familiar with all aspects of personal lines insurance, therefore, they would be incapable of exploiting the virtues of the on-line system. The project team stressed the importance to all underwriters of both avoiding inputting delays and using the system to deal with client enquiries without continually interrupting the work of more experienced staff by seeking advice and help. It was therefore proposed that the career path should be developed within personal insurance, bringing it up to a par with other areas of insurance. To this end it was proposed that there should be a widespread upgrading of staff.

Several key members of the project team had been convinced of the need to upgrade personal lines staff from quite early in the development process. They did not make this view public, however, until just before the major part of personal insurance work was to be put on-line. This delay was caused by the fear that early statements of this view would lead to strong union pressure for pay increases before management was in a position to assess the effectiveness of the new system.

Union reaction and the process of negotiation

In the preceding discussion we have simplified a long and complex process. Moreover, we have focused upon the development of management thinking. We now turn to a consideration of the role which the union played. The activities of the union in relation to the new technology can be conveniently categorized under a number of headings.

First, the union tried to ensure that it was informed about, and later more fully involved in, the implementation and monitoring of schemes. Hence, on a number of occasions, management plans were delayed because the union objected to a lack of consultation (which was required according to the procedure agreement). Subsequenty, the union won management agreement to the principle that local union representatives (reps) should be involved in the implementation of on-line system developments at area level. In fact, the extent to which this actually happened appears to have been limited, largely due to variations in the coverage and role definition of reps. Hence in our survey only a third of the reps said they had been involved in local discussions with management over new systems.

The union at company level also sought to monitor the changes by asking reps about their experiences at local level. The efficacy of this tactic was initially somewhat limited due to the failure of the reps to reply in much detail, or indeed at all. Accordingly, when routine amendments were shifted to the subsidiary's processing unit, the division asked specific reps to monitor the changes more closely, and specified the issues on which they were to obtain information. In this sense there appears to have been an increasing union role, both in terms of its recognition by management and the union's use of its own resources. However, there were still complaints on the part of the union that there were frequently lapses in the consultation process, and that they were consulted only at the stage that plans were to be piloted or implemented, rather than in the earlier development of ideas.

Second, the union initially appeared to be thinking in terms of a grand strategy in relation to new technology. In the late 1970s and 1980, and in part due to advice from the national union, a number of sub-committees were set up by the division to consider VDU developments and subsequently the more general question of the implications of new technology within the company. Some of the union activists were employed in jobs which intimately involved them in the applications of new technology. One role which the union committees played, therefore, was to permit these experts to advise the union. The union's deliberations led to the demand for a new technology agreement. Management rejected this on the grounds that it did not wish to foster a 'new technology phobia' among staff. However, the demand was one factor leading to the conclusion of a set of more general agreements concerning, for example, redeployment and redundancy, as well as jointly agreed guide-lines concerning the use of VDUs.

Much of the concern of the union came to centre upon the health and safety aspects of using VDUs, and this led to the guide-lines concerning their use. In order to achieve this agreement, the union had to make considerable use of its local reps to collate evidence that, contrary to management's initial claims, some staff were working on VDUs for extensive periods of time and therefore might be encountering eye-strain. However, even when the guide-lines were agreed it became evident that their application was seen as optional by some local managers. The union had, therefore, to insist that local management be informed that the guide-lines were mandatory. In addition, the union pressed for acoustic covers in certain situations, and was again finally successful.

Other aspects of union strategy in relation to on-line processing merged with more general questions which were developing at this time. For example, the on-line system created the possibility of redundancy and redeployment. But similar issues were arising as a result of a range of other long- and short-term policies which management was adopting. The union became increasingly concerned about management's staffing plans. It sought, and finally gained, information from management concerning job losses, and then proceeded to negotiate about these and how any surplus staff were to be dealt with. It set up its own sub-committee on restructuring and employment. However, the work of this body was soon superseded by events, as management introduced new proposals on staffing levels. But it did develop a set of ground-rules on staffing and workloads which were accepted by management, although they did not differ significantly from the way in which management was planning to act anyway (certainly there was no argument over them and they were not subsequently referred to). Over a number of years of discussion and negotiation—and for the first time in the company—agreements were made concerning redeployment and redundancy. But the union had to resort to a ballot, which supported strike action, in order to get movement in the later stages of negotiations on this and a number of other issues.

Problems of redundancy and redeployment directly attributable to on-line processing, however, were relatively minor before our research. A threatened enforced redundancy, challenged by the union, subsequently turned out to be unnecessary due to higher than expected turnover. The question also arose of whether the newly graded jobs should be advertised as new jobs; initial disagreements over this were overcome by a compromise. However, the union did become increasingly concerned about the scale of potential job loss—much of it occurring through the annual budget system—and increasingly questioned the employment implications of any new pilots or plans of management. Hence, for example, the central processing of personal endorsements, which was introduced during our research, led both to close questioning of management as to whether this was part of a longer-term strategy and to the more detailed monitoring mentioned above. The union was particularly concerned that this might be the beginning of a trend towards a centralization of personal insurance work, leading to a loss of jobs and skills.

A further theme running through the union's tactics concerned pay and effort. The latter was closely related to the question of job loss: in other words, the union was not merely worried about job loss but also about the implications of a reduced staff having to handle the same or greater amounts of work. In the early stages of the shift to on-line systems, for example, the union did take up with management the fact that manning levels—as a result of earlier job reductions through the budget—were insufficient to handle the workload. Subsequently, manning levels were increased as management realized that this was indeed the case in a number of areas.

The union was also involved in the grading of the jobs associated with the new on-line system. Management's proposals concerning the upgrading of personal insurance staff were clearly looked upon with favour by those staff most directly involved. But the union leadership feared that some other union members might oppose these proposals on the grounds that their status was thereby demeaned. The union, in the words of one activist, 'took the coward's way out and agreed that the new jobs should go to the job evaluation committee'. The union did, however, question the gradings attached to a number of the new jobs, seeking to get them upgraded. More recently, it feared that the shifting of work to the subsidary's processing unit was an attempt to substitute cheaper subsidiary staff for higher-paid underwriters. The union therefore insisted that the staff concerned should have particular gradings and levels of pay. Management gave them reassurances on these matters.

It would be wrong to suggest that the union began to develop any strong opposition to technical change as such. It recognized that this, and associated job loss, might be the means of improving the viability of the company and hence preserving larger numbers of jobs in the longer term. But, equally, it was afraid that management would see cost-cutting, and particularly a reduction of labour costs, as the only way to achieve profitability. It increasingly argued that there were means whereby revenue might be increased, and that costs other than labour costs might be reduced. These arguments were put forward only in general terms, and were not developed into any coherent policy proposal (although criticisms were made of what was deemed to be profligate expenditure on the part of certain managers).

Overall, the role of the union in the process of technical change can be seen as consisting of three elements. The first was the establishment of its own role in the process of planning and implementation. The second was to provide a series of national or company level guide-lines concerning key aspects of the change process. The third—which sometimes led to the second and sometimes derived from it—was to take up detailed points and problems, often concerning only one or two individuals, which were passed up to national level. They achieved some success on all three counts; for example, they did become somewhat more involved. But they achieved relatively little influence over work organization. Further-more, the gains achieved were often partly attributable to the views of certain managers, although these were shaped by an awareness of possible union reactions. One important reason for the limited role of the union was the weakness of union organization at local level, where the amount of bargaining was typically limited. Only a minority of local reps discussed the new systems with local management; less than half,

according to our survey, made much use of the VDU guide-
lines.

STRUCTURES OF LABOUR REGULATION AND UNION INFLUENCE

In discussing the labour aspects of the move to on-line
processing we have shown that there were a series of broader
changes within the company which were associated with some
expansion in the role of the union. Significant though these
developments were, it still remained the case that the union
achieved only a limited influence over matters relating to new
technology. This relative weakness reflected the more general
limitations upon the role of the union in the company. These
were related in large part to the broader structure of labour
regulation, with its emphasis upon individualism. In this
sense, the situation at the company was similar to that found
in many white-collar situations. In this section we detail the
nature of the system of labour regulation and the limited role
which the union played. In the next section, where we turn to
the question of union organization, we shall also consider a
variety of changes which were leading to a somewhat greater
role for the union.

Traditionally insurance companies have operated strong
internal labour markets: that is, there have been a limited
number of ports of entry into the firm and these have typically
been at the lower levels. Vacancies for higher-grade jobs are
filled internally. This was true of the company studied. The
union, however, had no influence over recruitment and
promotion procedures, nor over the criteria to be employed,
nor over whom management selected for vacant posts.
Nevertheless, the union was strongly committed to the
maintenance of the internal labour market which, from the
viewpoint of members, constituted a career structure. Hence,
for example, on a number of occasions the union questioned
why management had chosen to recruit staff from outside the
company; this was typically explained in terms of a lack of
internal applicants. In another instance, the union questioned
why a number of temporary staff had been recruited;
management justified this course of action in terms of

the short-term needs of a particular department during its transfer to a new location. The union accepted this explanation. At local level union reps played little role in matters of recruitment and promotion. Only 5 per cent of the reps replying to our survey said that they 'often' dealt with these matters, and well over half said that they never did so.

However, the internal labour market was becoming modified in two ways. First, with computerization the company had growing needs for those with the relevant skills; these were in short supply within the company, and hence external recruitment was significant in some areas. Second, in a number of peripheral areas, such as catering, management had dispensed with its own staff and had employed the services of a contractor. This was done without the agreement of the union.

As has been noted in the previous section, the union did begin to expand its role from the mid-1970s over matters concerning redundancy, redeployment, and effort levels. A new procedure agreement required that management discuss such matters with the union, and in this way the union did achieve some influence. For example, in one case it was able to 'save' seven jobs, but eighty redundancies were still made, almost half of these being compulsory. In another case, management initially rejected the union claim that proposed staffing reductions would lead to excessive workloads. It was only when subsequent experience with reduced staffing levels made it evident that the union's initial argument was correct that management increased the number of staff.

The extent to which reps discussed staffing and effort levels with local management was generally limited. In the survey only one of the respondents said that he always did so, and a further 13 per cent said they often discussed these matters with management. Half the respondents, however, claimed that they did so only 'sometimes', while a third stated that they never did so. Where reps were so involved, it appears generally to be attributable to members complaining about their workload to the rep.

In addition to seeking to influence the scale of job reduction and staffing levels, the union would also take up the cases of individual members. This frequently occurred at national level. For example, on one occasion the union successfully objected to a proposal to make certain staff redundant, claiming that management was using the redundancy procedure to 'settle old scores' or to escape the disciplinary procedure. More active local reps reported that they also represented members over such matters, and in some cases were able to achieve local deals with management, which, for example, ensured that a pregnant woman, who was in any event planning to leave, was included among those to be made redundant.

Union influence was rather more in evidence in relation to the terms and conditions of redundancy and redeployment. In addition to specifying the level of redundancy pay, the redundancy agreement detailed the steps to be adopted in the case of job reductions, with compulsory redundancy being the last stage (after such moves as the suspension of overtime and recruitment, and seeking volunteers for redundancy). Jointly agreed guide-lines concerning the redeployment of staff included a guarantee that salaries would be maintained, and specified the financial assistance which would be provided by the company where a move to another location was required. The union took up individual cases within the context of these agreements.

Beyond questions of redeployment and redundancy and seeking to ensure the maintenace of the traditional practices associated with the internal labour market, the union played very little role in shaping the structure of promotion and the allocation of jobs. Similarly, it played no formal role in the allocation of work within the office. In discussing work organization before and after the move to on-line processing, we noted that the superintendent allocated jobs between members of staff. This was done without any consultation with local representatives. In principle, however, it was possible for members to complain to their local representatives about such matters. In our survey of union reps, we asked how often they dealt with issues concerning job allocation: none said that they did so frequently; just under a third said that they did so 'sometimes'; but over two-thirds said that they never became involved in job allocation matters.

Another aspect of job allocation concerned workers of a

lower grade performing the tasks of a higher-graded job when the latter was vacant due to absence or turnover. Again, the union generally played no role here, although early in the 1980s it did seek an agreement with management concerning the pay of such temporarily upgraded workers. Traditionally, management had argued that such deputizing provided the worker with the opportunity to expand his skills, thereby improving promotion prospects, and that such co-operation with management would also be reflected in merit payments. Management did, however, accept the need to lay down guide-lines (rather than a formal agreement) on this matter, but still insisted that payments should be subject to management discretion and would only be paid after a fairly lengthy period of undertaking the higher-grade work.

The union had no influence upon the level or allocation of overtime, although it did negotiate overtime rates as well as holidays and hours of work. It had agreed a two-year experiment with flexible working hours in the late 1970s, and, at the time of the research, was involved in seeking to amend the scheme and extend it to the subsidiary company. Locally, reps played a limited role in dealing with the details of holiday and working time arrangements: less than one in ten said that they were 'always' or 'often' involved in such matters, while two in ten said that they never were. The union also played no role as far as the training of staff was concerned, other than negotiating the payment of certain expenses to those who undertook the relevant professional examinations.

The union had the right to deal with disciplinary issues according to the procedure agreement. Occasionally items of this kind arose at national level, but it seems that disciplinary issues were typically of limited importance as such. More important was the way in which management could affect staff through job allocation, promotion, and the payment of merit increases.

Members' earnings were made up of a number of components. The most important element was the basic pay of the grade in which they were employed. Actual pay rates were negotiated annually at company level, while grading was determined by a job evaluation scheme in which the key components were 'know-how', problem-solving, and accountability. In all, there were seven grades, above which more senior staff were paid according to the points which their job achieved in job evaluation. Jobs were evaluated by a joint committee, the union members of which had been specially trained. Gradings were made largely through consensus, although in a few instances disagreements were taken up within the main negotiating and consultative procedures. Given the scale of the changes which were occurring within the company, a considerable number of jobs had been regraded since the late 1970s—those involved with personal lines insurance being one good example. While the union was involved centrally in job evaluation, local reps typically had little role to play in job evaluation and grading. In the survey only two respondents said that they were always involved, and a further 8 per cent said they often were. However, nearly half said that they never dealt with such issues.

For each job grade there was a salary range. Movement up the range was based upon management assessments of merit, although some annual or biennial increase was the norm. As has already been noted, the union rarely became involved in questions of merit increases, and a number of senior reps expressed the view that it never should. However, the union was keen to preserve the principle of incremental merit awards, and in one case challenged management, claiming that it had broken the convention of paying them. This was one of the factors which led to the ballot for strike action: management finally agreed to remind local managers of the need to pay merit awards.

Some groups within the company, such as those involved with the sale of policies, were also paid bonuses based on sales or some such criterion. Such payments varied in importance; for one group, for example, they averaged about 7 per cent of pay, although individual bonuses varied between a low of 2 per cent and a high of 32 per cent. The union discussed such bonuses and the details of their payment with management. In addition, they were involved in the establishment and continuation of trial bonus schemes.

A wide range of other payments and fringe benefits were paid to staff. These had constituted an important topic of joint discussion in the days of the staff association, and, while relatively less significant in the 1980s, they continued to be negotiated and extended. As well as maternity and paternity leave, pensions, and sick pay, fringe benefits included London and other regional allowances, the provision of company cars and car allowances, clothing allowances, help with house purchases, and various favourable terms for the purchase of insurance. The union was also represented on the pension fund trustees, and in the early 1980s was thinking of seeking parity representation on this body; this demand was allowed to lapse, however, given the range of other issues which were being dealt with. In addition, there was a staff share scheme, although management was loath to permit the unions any influence over this. For example, the union had demanded that part-time staff be eligible to join the scheme, but management rejected this on the grounds not only that they were not prepared to make this an issue of negotiation, but also that it was designed solely for those who thought in career terms, and this was said not to be true of part-time staff.

A network of health and safety committees existed within the company, and such issues were also the subject of discussion at national level. It has already been seen that this formed an important theme in the discussion of the move to on-line processing. In addition, it appears to be the single most common area of activity on the part of local reps. For example, over a third of them said in the survey that they always or often deal with such issues, and a further 50 per cent said that they sometimes did so. Again, these issues were generally of a relatively minor nature, such as requesting that the heating system be repaired or switched on.

It is clear from the preceding discussion that negotiations and consultation were highly centralized. The most important formal joint bodies were the joint co-ordinating committee and the JNC. The former body, upon which sat the division secretary, chairman, and two other members of the divisional committee, along with an equal number of senior managers, met at least monthly and discussed a wide range of issues. It was here that the unions were informed of management plans, including details of staffing levels and annual budget proposals, and many more detailed issues were raised. Since the shift from a staff association to a trade union there had been a change in the items of discussion from individual and welfare matters to more conventional industrial relations issues. In part, however, this was attributable to the scale of the changes which were occurring within the company. Many issues which in other companies would have been the subject of formal negotiation—such as various aspects of the wage–effort bargain—were the subject of discussion in this essentially consultative body. In addition, a joint forum was held every six months, where all senior union representatives met a delegation of more senior managers to consider the financial state of the business. This body also provided the union with an opportunity to obtain company information on a wide range of matters.

A few times a year the co-ordinating committee was replaced by the negotiating committee, consisting of the same delegations. This dealt in particular with the annual wage claim. More generally, as the preceding paragraph suggested, there was little clear demarcation between consultation and negotiation. This was reflected in the procedure agreement, which declared that 'the company agrees that before implementing any innovation involving alteration to matters covered by the Scope of the Agreement it will notify and if necessary consult the Union and resolve any consequent issues through the procedures laid down by the Agreement.' Most of those procedures involved essentially consultative bodies. Nevertheless, it is clear that when matters were raised in the co-ordinating committee management was typically loath to go ahead with its plans, without some degree of consensus being obtained with the union. However, the union was often concerned that issues failed to be brought early enough, or at all, to this body.

At area and head office level, joint local committees existed, in which local union representatives discussed matters with management. These typically met about three times a year and appear generally to have been of limited significance, given the high degree of centralization within the company and the limited bargaining awareness of members and local reps. As already noted, there were also health and safety committees. However, as is indicated by the preceding paragraphs, the range of local discussion was generally limited, although it varied considerably between offices. The extent of local bargaining depended upon the nature of the rep and management: it was not related, according to our survey findings, to the size of office, the level of union membership, or the precise nature of the office. The vast majority of reps—over four-fifths—spent less than one hour per week in discussions with local management. Nevertheless, the majority did claim that they could resolve most issues locally, without resort to a higher level of the union. This appears, however, to reflect less the existence of strong bargaining relationships or influence than the nature of the issues typically raised. Most appeared to be minor matters which were not contentious. When disagreements did arise, the division secretary and the central personnel manager quickly became involved. Indeed, a crucial feature of the pattern of industrial relations within the company was the strong bargaining relationship which existed between these two. Both were pivotal figures for their respective organizations, and both encountered a considerable degree of uncertainty over the reactions of staff and many managers (who were still finding it difficult to adjust to a style of management which recognized the existence of a trade union). The strong bargaining relationship therefore permitted the sounding out of issues and the formulation of mutually acceptable compromises.

In overall terms, then, the range of union influence was growing. This was largely attributable to the overlapping of the merger of the staff association with BIFU and the scale of change which was occurring within the company. Union influence, however, was concentrated at the level of the company, while locally the range and significance of bargaining were both variable and limited. Multi-level bargaining was very rare, despite the fact that the framework of agreements and understandings provided a base for such activity. In part because of this weakness, the union's influence was limited to setting down certain parameters within which management action could take place, rather than involving the union in the detailed regulation of work organization. Moreover, while it appeared to be of little concern to the union, it might be suggested that the relative weakness of the union was indicated by the fact that much of its influence rested upon an essentially consultative rather than negotiative process.

UNION ORGANIZATION

In the preceding sections two aspects of union organization are worthy of particular note. The first of these is the way in which union influence was growing. The second is that at company level the union organization appeared to be relatively active, while at office level it was far less so. This section looks at these themes more fully, and goes on to consider the nature of staff attitudes towards the union.

The development of trade union organization

While the growing uncertainty of staff, as rationalization proceeded, was a factor encouraging the expansion of the union's role (cf. Prandy *et al.* 1983: 147), changes in union organization were also important. A staff association had existed in the company for many years and had been registered as an independent organization under the 1971 Industrial Relations Act. The association had merged with BIFU (known at the time as NUBE) in the latter part of the 1970s. The stimulus to seeking such affiliation was largely a chance factor—the take-over of a smaller company organized by ASTMS, so that it was necessary to come to some agreement concerning representation. However, there were also other, longer-standing factors. The resources of the staff association were becoming stretched, and it was increasingly felt that affiliation with a larger organization would relieve this problem and give the association greater strength and independence in its dealings with the company. Discussions had been held with similar staff associations with the intention of forming an association of insurance staff associations, but these had broken down. Approaches were therefore made to three trade unions to discuss possible mergers. The staff association finally chose BIFU, partly because it offered the best terms, e.g. a significant degree of autonomy as a division within the union. However, other considerations were also important and reflected the notion of trade unionism held by the members and activists. One attraction of NUBE was that it was an industrial union catering solely for the finance sector, and this was seen to be important since it indicated an awareness of the distinctive problems and interests of members. Second, NUBE was not affiliated to the Labour Party, and it was deemed important to avoid any political connections.

The transfer to BIFU was associated with a growth of union sophistication. First, union membership grew. In 1978 union density was only 55 per cent, but had reached 85 per cent by 1983—a high level for the finance sector. This expansion largely occurred among lower grades. The staff association's membership came largely from middle and higher grades, although these continued to dominate the official positions within the union (this was a factor explaining why only a quarter of reps were women, despite the fact that they made up a much larger proportion of the labour force). Second, union organization within the company became more firmly rooted: a network of local union representatives developed (albeit incompletely); structures were introduced which served to link local and company (or division) levels of organization through area and regional bodies; and different interests became more fully represented at division level. The staff association had had virtually no organization below company level, and there had been very little contact between members and company-level bodies, beyond an annual conference. Third, organization became more developed at company and division level. The staff association had an executive committee but the bulk of the work was done by its general secretary. This pattern still continued to some degree with BIFU, although a structure of sub-committees and working parties was developed which provided greater support for the full-time secretary and permitted the involvement of a larger number of activists. Fourth, the staff association had been an isolated, independent body unable to obtain information, advice, or support in any consistent and coherent manner from other sources. Membership of BIFU dramatically changed this situation. The larger union became an important source of support for the division. As a result of these changes, the role of the union expanded. The staff association had devoted a great deal of its energies to its own internal administration and, in its dealings with management, was mainly concerned with a variety of pay-related issues and fringe benefits. The move to BIFU was associated with a growth of formal agreements and undertakings, establishing procedures, and dealing with a wider range of issues.

While the changes occurring within the company were a factor in the increase in union density, the union itself played an important part in encouraging membership and expanding its role. If a variety of other factors had not led to an association with BIFU, it is open to question whether the degree of joint regulation would have expanded to the extent it had. There are several reasons for this. First, the personnel function in the company was very aware of the need to change its approach to labour relations since the demise of the staff association. Second, the increased sophistication of the organization within the company was a crucial factor in expanding the union's role. Third, the growth of a collective orientation on the part of the membership, conditional though it was, was to a large degree dependent upon the activity of the union in fostering an awareness of the role which it might play and demonstrating its ability to promote and defend members' interests. (The growth of unionization in this case indicates the need to integrate more fully the differing approaches which have been adopted to white-collar unionism, as seen in the work of e.g. Blackburn and Prandy 1965; Crompton 1976; Bain 1970; Bain *et al.* 1973. On unionization in insurance: see Crompton 1979 and Heritage 1980.) Nevertheless, union organization still remained relatively weak in a number of key respects. In pursuing this point, we can look at the various dimensions of union sophistication.

Union sophistication

It is convenient to consider the structure of the union within the company by starting with organization at company level. Here the union was very sophisticated. As a relatively autonomous division within BIFU, it held its own annual delegate conference, had its own executive committee and a variety of other committees which met regularly, and a full-time secretary.

The key decision-making body was the annual division conference. This met for two days in working time, and involved about fifty specially elected regional and area representatives, as well as division officers. It passed quite detailed resolutions on a wide variety of issues, and was not dominated by the division committee. The latter was the key executive body, playing a central role in all union matters within the company. It consisted of eleven representatives elected by the regional committees, plus three officers—the chairman, deputy-chairman, and treasurer—elected by the annual division conference. In addition, the key figure on the committee—the division secretary—was appointed by the division committee in co-operation with the general secretary and president of BIFU. Union representatives on the various joint committees with management—such as the joint co-ordinating committee, the JNC, the job evaluation committee, and various working parties—were selected by the division committee. The committee met at least monthly in working hours, and from time to time established special sub-committees to consider various issues in depth. The full-time division secretary spent a great deal of time in informal negotiations with management, as well as being the centre of a considerable network of union activists. In addition to his visiting local areas, it was common for local reps to contact him over relatively minor issues. At division level, therefore, the union was highly sophisticated; in shop steward terms, it had a full-time convenor, a body of senior shop stewards, and regular meetings of those stewards.

Not only was union organization at division level sophisticated, but it was also strongly integrated with the larger union. The division committee elected representatives to BIFU's insurance section council and, for most years since the transfer to the BIFU, a representative from the company had sat on the union's national executive. Beyond these formal links at company level, there were frequent contacts with the larger union. These often involved seeking ratification of the division's actions and policies from the larger union, as well as a two-way flow of information. In particular, the division sought information from the union's research department and national officials on a variety of issues, e.g. new technology, as well as receiving the general flow of documents on union policy. The division secretary also represented the larger union on a number of bodies, while national officials attended the annual division conference, and sometimes discussed problems with the division committee.

It is also convenient to mention briefly here the existence of a company managers' association, representing grades above assistant manager level. This body had existed for a number of years, although it was not recognized by management for collective bargaining purposes. Its primary function was as a channel through which local management views on company proposals could be put to central management. Some members of this body were also members of the union, which, towards the end of our research, was considering the possibility of establishing a separate management section within the division. The division had no formal links with the association.

At division level, then, union organization was sophisticated and strongly integrated with the larger union. Its weaknesses lay below this level, despite a structure which sought to provide linkages between the centre and the office floor. Below company or division level there was a network of regional and area committees. There were seven regional committees covering from two to five area offices and their branch offices, plus three head office committees. Although the latter typically met more frequently than their regional counterparts (whose members were often widely dispersed geographically), they played only a limited role. Regional committees typically met only once or twice a year, although some contact between activists within regions was maintained by telephone. The size of the committees depended upon the size of the membership, and generally totalled four to five members, these being elected by the area committees. There were twenty-six of the latter at the level of the company's area offices, typically having four or five members. The frequency with which they met varied; about a third met once a quarter, but an equivalent number met only once a year or not at all. Again, the infrequency of meetings reflected their limited role. In practice, none of these regional or area committees were involved in negotiations, nor did they constitute a stage in the internal union procedure for dealing with industrial relations problems. In some cases they provided a forum for the exchange of information. Their primary significance lay in selecting representatives for company-level bodies and in formulating resolutions and views which were then passed on to the division committee or put forward at the annual conference. In terms of the co-ordination between different levels of the union, therefore, there were significant weaknesses: the structure existed, but did little. There was no equivalent of a regular meeting of shop stewards.

The weakness of organization below the company level was reflected not only in the limited role which regional and area committees played, but also in the structure and role of local reps. The previous section showed that they were seldom involved in office-level negotiations. Few spent more than one or two hours per week on union business, and much of this time was spent talking to members or other union reps. A third of reps reported that they held no meetings of members, while only a third said that they held such meetings more than twice a year. Moreover, where such meetings were held, reps claimed that generally less than a quarter of the membership bothered to attend.

At local level, the links with the larger union, or indeed BIFU reps outside the company, were weak. While reps claimed that almost half the membership read the union journal and other union publications, very few members or reps had any contact with BIFU locally. Only two representatives from the company sat on BIFU area councils, while two-thirds of reps said they had no contact with BIFU reps from other companies; only two said they had such contact monthly. The pattern of external integration, then, reflected the pattern of internal sophistication—high at company level, but weak at lower levels of the organization. This was attributable to the pattern of 'shop-floor' organization.

Many local rep positions to which the union was entitled under an agreement with management remained unfilled. In 1983, for example, only 99 of the total possible number of 143 were filled—that is, 69 per cent. This was a somewhat lower proportion than a few years earlier, when 75 per cent of posts were filled. A detailed analysis of the pattern of rep vacancies indicates that there is no relationship between union density and the occupancy of rep positions. More important is the difference between area offices and central units. In the early 1980s the proportion of positions filled varied little between these. However, at the last elections (these were held every two years) the proportion of posts filled in areas rose from three-quarters to 90 per cent. In the central units the proportion fell from about three-quarters to about 45 per cent. Within the central units there was no clear pattern of vacant positions by the type of function performed in different constituencies.

Three factors appear to be relevant in explaining this variation in rep occupancy rates between areas and the central units. The first concerns the differential impact of rationalization. In the area offices there was a good deal of concern over the changes being introduced by management, which encouraged resort to the union, but there was little redeployment of staff. Hence, members formed relatively stable groups. The situation was rather different in the central units. While there was a similar concern over the effects of rationalization, there had been considerable redeployment of staff and relocation of departments. This meant that there was rather less stability within groups, and that potential reps may have been preoccupied with personal problems concerning moving home and resettling their families. A second factor appears to be the contrasting approaches of management. Management in the area offices was fairly tightly controlled by the central personnel function, and therefore demonstrated little hostility to union reps. It seems that the situation was rather different in some of the central units. Here managers had somewhat greater autonomy, and according to reps, a number made clear their hostility to trade union activity which spread beyond membership. This approach on the part of management was especially effective in discouraging members from becoming reps, because staff in the central units tended to be more ambitious than their counterparts in the area offices.

However, underlying this pattern of incomplete representation is the basic fact that union membership was incomplete and that, even where it was relatively high, there was not a tradition of trade unionism which induced groups of staff to ensure that local reps were selected. The existence of union reps often simply reflected the readiness of individuals to stand, so that the pattern of union representation was in part determined by chance. Rep positions were rarely contested— only five were in the most recent elections. On the other hand, where a number of candidates stood and an election was held, the turnout always exceeded 80 per cent of the membership.

Member ambivalence and organization

Fundamental weaknesses of the union, therefore, lay both in incomplete membership and also in the attitude towards the role of the union on the part of those who were members. Interviews with union reps, members, and managers all indicated that the definition of trade unionism typically held by members was ambivalent and uncertain. At the extreme was the hearty independence of some non-members who declared 'I can fight my own battles'. More generally, most members—including the vast majority of reps according to our survey—saw the union as a form of protection, like the insurance policies they dealt with, rather than a movement which required their active commitment.

The limited commitment to collective methods reflected the nature of the work and market situations of employees, which fostered an individualistic approach (cf. Lockwood 1958). Four factors were of particular significance in this respect. The first of these was the existence of a career structure which, given the importance of internal labour markets both in the company and in other parts of the finance sector where their skills were marketable, fostered identification and co-operation with the company as the optimal route to increased earnings and other valued rewards from work (cf. Batstone 1984 on the significance, in the Japanese context, of most employers operating internal labour markets for particular occupational groups). The primary means of self-improvement, therefore, were of an individual, rather than collective, nature.

Second, such individual identification with the company

was further encouraged by the traditional personnel policies pursued by the company. These had always been of a highly paternalistic nature, involving a wide range of fringe benefits, as noted earlier, and the provision of a wide range of company-based social activities. In the past, such paternalism had also been characterized by a fairly rigid exercise of authority relating to matters of dress and decorum (see e.g. Supple 1970). But such requirements had served to heighten the status consciousness and sense of distinctiveness among staff. These features had become a good deal weaker over the last decade or so, but they still figured importantly in the thinking of many longer-serving staff.

Third, many of the staff were undertaking tasks which, to varying degrees, were managerial in nature (cf. Carchedi 1977). While the discretion of underwriters in personal insurance may have been declining over the years, some still remained, particularly in commercial insurance. In effect, therefore, the performance of day-to-day tasks fostered identification with the company. This characteristic of work was even truer, of course, in relation to the substantial numbers of supervisors and functional specialists who were members of the union. Thus, there could not easily develop a rigid 'them and us' conception of relationships within the company.

The fourth factor concerned the nature of day-to-day relationships within the company. In addition to the way in which career routes and authority structures overlapped (this was discussed in relation to the chemical case-study, but was even more true in the insurance company), work units were typically small, with easy and informal relationships between routine grades and lower levels of management. Typically, therefore, few difficulties arose if staff wished to raise problems and grievances with their superiors. Indeed, to seek the aid of the rep might often have indicated a lack of trust between the individual and management, which would serve to disrupt the more general pattern of relationships.

Nevertheless, union membership had grown substantially and on one occasion, a few years before our research, the membership had voted in favour of strike action. These developments can in part be seen as attributable to the role of the union itself, a theme to which we return below. However, the thrust towards a somewhat more 'unionate' approach on the part of staff lay in the fact that the rewards of individualistic strategies appeared to be under threat. This was particularly true as far as the uncertainties arising from rationalization and cost-cutting exercises were concerned. It also arose from wage increases which failed to preserve the traditional status of the staff and from a series of other incidents which could be interpreted as indicating 'bad faith' on the part of management. Issues of all these kinds led to support in the ballot for strike action.

Collectivism, therefore, grew out of the dangers which threatened the traditional individualistic strategies of the staff (it is not to be explained, therefore, simply in terms of proletarianization, at least as conventionally discussed—cf. Hyman and Price 1983: Pt. II). In many respects, members defined the role of the union as defending and preserving the gains which they had come to expect from such strategies. Hence, the role of the union was seen not so much in terms of becoming involved in promotion decisions or in the allocation of merit payments, but rather in protecting career structures and the internal labour market. The union's task, as many members saw it, was to guard the parameters and 'rules of the game' and to ensure that conformity 'paid off', rather than to become directly involved in playing the game itself.

Members' views of the proper role of the union and a centralized management structure therefore led to a concentration of union activity at division or company level. However, the role which the union could play, even centrally, was seriously constrained by the weaknesses of organization and orientation at office level. The division activists were seeking, through a variety of means, to overcome these weaknesses—and with some success. They sought to maintain links with the membership in a number of ways. Important among these was the distribution of a union newsletter. A dozen or so of these were produced each year, typically outlining major issues, the reasons underlying the adoption of a particular approach on the part of the union, and the stage of negotiations. In addition, visits to local offices were made by the division secretary and chairman, while it was common for individual members to contact the division secretary over personal problems and grievances.

The role which the union was able to play was constrained, ultimately, by the nature of members' attitudes. In particular, it was forced to adopt what it termed a 'responsible and reasonable approach' since in its actions it was often seeking to persuade its own members as much as management. It had to recognize the commitment which many of the staff felt towards the company. Such considerations are to be seen clearly not only in the detailed arguments of the union in negotiations, but also in a union pamphlet at the time that strike action was threatened. This document continually referred to the emphasis which the union had always placed upon 'responsible trade unionism' and the importance of 'harmonious staff relations'. Hence, the document claimed, management 'have not been faced with staff representatives who were obstructive as a matter of principle and who backed up the principle with threats of sanctions against the Company'. The union's actions, it was argued, 'have always been moderate and patient' and the move towards strike action was attributed directly to the failures of management— 'responsible trade unionism needs responsive Management . . . if responsible behaviour is frequently frustrated or is mistaken by Management for weakness then the outcome is inevitable—conflict.' The pamphlet continually stressed that strike action had only been considered as a last resort and because 'Management left us no acceptable alternative'. It concluded with the statement that a change towards a conflict-ridden pattern of industrial relations was an appalling prospect from the point of view of the union and of 'the large number of staff to whom the company is more than "just a job" and whose proven loyalty deserves much better than the treatment which they are receiving currently'.

At the same time, the growth of union density and the vote in favour of strike action (55 per cent, with a turnout of 80 per cent) meant that management could not simply ignore the union. There was the possibility that, by trying to do so, management would not only strengthen the commitment of staff to the union but also endanger the traditional loyalty of the staff to the company—something which the company had carefully fostered. As a result, the personnel function played a very central role within the company. All proposals which had any implications for staff relations were carefully monitored by the personnel manager, who also operated a firm control over labour relations issues at local level.

The nature of staff and member views and their significance for both the company and the union therefore produced a situation of uncertainty for both parties. This made the strong bargaining relationship between the personnel manager and the division secretary of central importance. The confusion between consultation and negotiation can also be seen as a direct reflection of this situation. These two features of labour relations permitted a good deal of flexibility and the opportunity to reach mutually satisfactory accommodations without entering into the uncertainties of continual appeals to the staff.

At company level, therefore, the union was highly sophisticated, but this was not the case at local level. Despite a formal structure designed to integrate the different levels of union organization, the significance of these links was limited, other than as a means of transmitting problems to the central level. In this way, the importance of the company level was reaffirmed. The limited degree of local sophistication was largely attributable to the nature of the members and their work situation. However, union organization had become stronger in recent years, due to the interaction of the transfer of membership to BIFU and related changes in union organization, and the membership's fears concerning the rationalization and cost-cutting initiatives of management.

In this chapter we have discussed the move to on-line processing of personal lines insurance in one company. This was part of a broader strategy aimed at increasing competitiveness and reducing administrative costs. Indeed, the nature of the broader strategy had a greater impact upon work experience, in terms of effort and employment levels, routinization, etc., than the new technology. The main significance of the latter was to reduce the amount of paperwork and eradicate the need for underwriters to do arithmetical calculations; it also limited the order in which various tasks could be done. As a result of these changes, the underwriting services section was integrated with the under-writing section. Many staff were upgraded, in part because it was deemed that greater knowledge and competence were required, and in part to improve morale and prevent opposition to the new system.

The role of the union in the process was largely confined to health and safety issues as far as work organization was concerned. It was also informed, and to some degree consulted, about broader aspects of management strategy. However, the job reduction and redeployment associated with management's more general strategies were the basis for an expansion of the range of union influence. This was also associated with a transfer of membership to BIFU, which in turn led to a number of organizational developments. These made the union more able to extend its role. Nevertheless, union organization remained relatively unsophisticated at local level. This was attributable to the ambivalent approach of the membership towards trade unionism. Such uncertainty shaped the approach of both management and union and reflected the paradoxes of using collective means to defend the opportunities and rewards of individualistic strategies.

6

CONCLUSIONS

In this final chapter we seek to draw together key findings and themes from the case-study chapters and relate them to a number of debates outlined in Chapter 1. In the first section, therefore, we look in some detail at the changes in work content and control associated with new technology, listing more systematically than in the discussion of the individual case-studies the various dimensions of task content. In doing so we consider more general arguments concerning the impact of new technology upon work and theories of technological determinism. The second section considers the question of the rationales underlying management's decision to introduce new technology, and why it was that the scale of change was typically fairly limited. Here we raise a number of general themes, in particular the extent to which technological innovation is shaped by a desire on the part of employers to control labour. The third section turns to the question of joint regulation and the role of the unions. Here we stress a variety of themes concerning the significance of union sophistication and highlight the importance of the interaction between union organization, member attitudes, and the nature of the work situation. We also discuss the impact of the recession upon the role of trade unions, and consider possible developments in their nature and role. The fourth and final section looks at the more general questions of approaches to the analysis of forms of labour regulation.

NEW TECHNOLOGY AND CHANGES IN WORK ORGANIZATION

As we noted in Chapter 1, there has been considerable discussion of the impact of new technology upon jobs. Views range from an optimism that demeaning, unpleasant jobs will be reduced in number to pessimistic predictions that new technology is yet another weapon in the armoury of employers seeking to degrade work. We pointed out that such arguments had a long pedigree and that much of the current debate in relation to new technology encounters problems similar to those in the more traditional debate. Important among these are the following: the differing focus of various commentators; insufficient attention to the various dimensions of work and related notions such as skill; and over-hasty generalization on the basis of less than adequate data.

In discussing our case-study findings we have concentrated upon the broad patterns of change in work content and organization. We focused on these, rather than providing a systematic analysis of jobs, in order to simplify the discussion and to highlight the pattern of change which needed to be accounted for. Much of each chapter was therefore devoted to such explanatory endeavours. However, we need now to look in greater detail and more systematically at the structure of work, in order to link our findings to the general themes which we outlined in Chapter 1. This is done in Table 3. As with our earlier discussion, we distinguish three broad areas. The first of these focuses upon the degree of worker influence over the production process: here we assess the extent to which workers can affect output, in terms of both volume and quality. We also consider the extent to which workers influence working methods, including the ability to vary the ordering and omission of tasks and cycle times. The next two dimensions concern what might be termed the conditions of competent performance. Under the general heading of the requirements of workers, we consider a range of factors which have important implications for such matters as opportunities for self-development, the level of job satisfaction, alienation, and the like. To some extent they overlap with the first dimension, but we have chosen to allow such overlapping because they focus upon different aspects of task performance. 'Sanctions and constraints' upon workers concern the various ways in which, consciously or otherwise, employers impose pressure upon workers to perform their tasks competently. These can be seen as various forms of control which operate at the task level, as distinct from more generally (e.g. through the existence of internal labour markets). Aspects of the broader pattern of labour regulation have not been included in Table 3 but are discussed below.

Before looking at Table 3 in detail, a number of points concerning its construction should be made. First, the scoring system, which is based upon observation, documentation, and interviews, is meant merely to be indicative. That is to say, we are concerned about highlighting changes and differences rather than measuring each dimension in any absolute sense. In addition, we have used a crude scoring system. We frequently felt inclined to increase the range of scores used in order to permit more subtle differentiations, but decided against this since in other cases we felt that the fine gradations required for more complex scoring were not justifiable in the light of the data we had obtained. Second, we have often averaged across a range of jobs, seeking to highlight the 'typical' pattern. This point is particularly important in, for example, the brewing and chemical case-studies, where there was a quite finely graded hierarchy. The effect of such generalization is, in the former case, to exaggerate the degree of change between the old and new breweries, since the scores for the former constitute an average across all grades. A similar bias occurs in the chemical case when we contrast the control room operator in the automated plant with the typical processman on the conventional plant. The third caution concerns the generalization and implicit weighting in each score. For example, much of the work of the chemical control room operator may be relatively routine, but in crises the complexity of his task is considerable. The score on this factor reflects the latter situation, which, it should be recognized, though exceptional, is extremely important. A similar problem of generalization is found in the case of the range of tasks undertaken by underwriters. Here, some changes—each underwriter being expected to handle all kinds of personal lines insurance, the fusion of underwriting and underwriting services, and the need to use a computer terminal—increased the range of tasks performed, while other changes—the reduction in the range of policies within any one area, the smaller scope for applying non-standard premiums, and the rarity with which staff were required to undertake arithmetical calculations—served to reduce the range of

TABLE 3. Dimensions of work organization: changes associated with the

	Chemicals		
	Processman (conventional dedicated)	Processman (automated)	Control room operator
1. *Worker influence over the production process:*			
(a) Output levels	3	2	4
(b) Quality	3	2	4
(c) Working methods (incl. initiation/ omission of tasks)	2	2	3
(d) Cycle times	2	1	3
2. *Requirements of workers:*			
(a) Effort levels	2	3	4
(b) Working in good conditions	2	2	4
(c) Manual dexterity	2	1.5	2
(d) Range of tasks	2	1.5	5
(e) Compexity of tasks	2	1	4
(f) Responsibility	3	2	5
(g) Ability to vary work pace	3	3	2
(h) Interaction	3	3	5
3. *Sanctions/constraints on workers:*			
(a) Output standards	4	4	4
(b) Quality standards	4	5	5
(c) Written operating rules	4	4	3
(d) Supervision/specialist checks	3	4	1
(e) Disciplinary action	3	3	2
(f) Technical stimuli/control	3	2	3
(g) Financial incentives	1	1	1

Note: Score: 1 = low, 5 = high.

tasks. The similar scores on range of tasks therefore conceal a variety of significant changes.

In general terms, Table 3 indicates that there are quite different tendencies associated with new technology, both between case-studies and between different occupational groups within some of the companies studied. Such variations would have been all the greater if we had included in our analysis other groups involved in the production process— maintenance staff, technicians, computer programmers, and other specialist managers. However, even if we confine our attention to the groups we studied systematically, there are marked variations. In the brewery, for instance, workers in introduction of new technology

Brewing		Engineering			Insurance		
Operator (old brewery)	Operator (new brewery)	Conventional machines	CNC lathe	Machining centre	Under-writer (old system)	Under-writing services clerk	On-line under-writer
3	3	4	4	4	4	4	4
3	3	4	4	4	4	4	2
2	4	4	3	5	4	4	3
2	4	4	3	5	4	4	4
2	3	2	3	3	4	4	5
2	3	2	3	3	5	5	4
3	2	4	3	4	1	3	2
2	4	3	4	5	3	2	3
2	3	4	3	5	3	2	2
2	4	3	3	4	2	2	2
2	3	4	3	3	4	4	4
4	4.5	3	3	3	3	3	3
3	3	3	3	3	2	2	2
3	4	4	4	4	3	3	4
3	2	1	2	1	3	3	4
4	1	3	2	2	2	2	2
2.5	2	1	1	1	1	1	1
3	3	1	3	1	1	1	2
1	1	3	1	1	2	2	2

the lager plant typically had greater influence over the production process on two counts—working methods and cycle times. At the other extreme, the move to on-line processing in insurance was associated with less worker influence on two key aspects—quality and working methods. Tighter underwriting rules and the fact that the computer now calculated premiums reduced the ability of underwriters to affect quality, while the requirement that data was input in a specified order reduced the scope for varying working methods.

In the other two cases, trends in worker influence varied between different groups of workers. In chemicals, it was already fairly limited on conventional dedicated plants. Contrasts in typical worker influence were more marked between multi-purpose plants and primarily manual production, on the one hand, and dedicated plants on the other. In addition, across all plants closer supervision and tighter operating procedures had served to reduce the scope of worker influence. Relative to these changes, the impact of automation in the case of the typical operator was limited. As we have noted above, the role of chargehands was somewhat greater, although only marginally so, on conventional plants. Even if the comparison is confined to these, the importance of the control room operator on the automated plant remains striking. His ability to influence production on all four counts distinguished was significant.

Variations in the impact of new technology are also found in the case of the engineering plant. In this case, workers on CNC lathes had rather less influence than their counterparts on conventional machines, as far as working methods and cycle times were concerned. Working methods were in part embodied in the programs and, in addition, the menu for programming acted as a further constraint upon variation. Once the CNC lathe had been programmed, the operator had little influence over cycle times. The situation was rather different in the case of CNC machining centres. Here, both the greater range of tasks which could be undertaken on the machine and the greater complexity of programming gave the operator a good deal more influence over working methods than was enjoyed by conventional operators. These factors, in turn, meant that those working on machining centres could have a far greater impact upon cycle times.

In summary, the first point to note is that the direction of change in worker influence over the production process varied considerably with the introduction of new technology. The scale of such change, however, was often relatively limited. The most dramatic changes, according to Table 3, are found in brewing and also for the control room operator in chemicals. Despite the fact that Table 3 exaggerates the scale of change in these two instances, this picture is broadly correct. It is, however, worth spelling out the reasons for the somewhat limited scale of change. As we have already noted, in several cases where worker influence fell it was already fairly low. This can be seen by looking at the conditions for competent performance. In the case of chemical operators the reductions in physical effort, manual dexterity, mental demands, etc., are all fairly marginal, so that the overall score falls by less than a fifth on requirements of workers. The same pattern is found, to some degree, in the case of underwriters. In other cases—notably CNC lathe operators as compared with conventional machine operators in engineering—the small changes reflect the continuing importance of worker influence over the production process. In this instance, less physical effort and manual dexterity were required of workers operating the new technology, and the simpler programming on CNC lathes reduced the complexity of the task. On the other hand, mental demands and the range of tasks rose somewhat due to the need for programming, albeit of a rather simple nature. But in these cases, and that of the underwriters, the limited impact of new technology upon worker influence and the requirements for competent worker performance reflect the fact that new technology directly encapsulated relatively few of the tasks which workers were still required to do. Hence, for example, a good deal of the operators' tasks in the engineering study involved conceptualizing what mechanical processes were involved in achieving the various end-products indicated by plans and instructions, and then setting the machine up to permit these tasks to be undertaken. While this was a somewhat more important task on conventional machines, it was still a key feature of the work even when CNC lathes were used. Conversely, where changes in worker influence and requirements for competent performance are greater, the worker spends considerably more of his time on tasks which involve him in working directly with the new technology.

In all the cases studied, although to markedly varying degrees, one key feature of new technology was that it assumed certain tasks which had previously been done by workers. Some such tasks were of an essentially routine, manual nature. In the brewing study, for example, much of the cleaning which was done manually in the old plant was done automatically in the lager brewery. In the insurance study, much of the essentially routine clerical work involved in writing policies was now done automatically. The same was true to some extent in the case of the CNC equipment in the engineering plant. But it is a second feature of new technology's assumption of tasks which has been the focus of a great deal of discussion. This concerns the way in which it assumes monitoring and controlling functions. Hence, for example, to varying degrees in both the automated chemical plant and the lager plant the monitoring and adjustment of temperatures and pressures which were otherwise undertaken by operators were now undertaken by the technology itself. Instead of gauges indicating the possible need for action on the part of workers, the new technology automatically undertook such regulatory action as might be required. Hence, ironically, far from new technology involving greater technical control over or stimulus towards action on the part of workers, it actually reduced it. Moreover, in those cases where workers operated control panels—as in chemicals and brewing—the nature of the relationship between man and machine was also distinctive. Certainly the control room operator received a series of indications of the need for action through the technology, and certain actions were precluded by the equipment, but the actions involved were often of a kind which required instructions to be given to other workers. Hence, technical stimuli fell for the typical operator in the automated chemical plant but did not for the control room operator. (The less sharp division between jobs in the brewing plant serves to obscure such contrasts—they would have existed if the original manning plans had been implemented.) Moreover, as a result of this change the operator in chemicals became more subjected to supervision: he was frequently instructed to undertake particular tasks. The control room operator, on the other hand, was subjected to less supervision.

However, an even more significant aspect of the work of the control room operators was that in many respects they monitored the monitoring equipment. They were able, in a variety of crucial situations, to use their own judgement rather than relying exclusively upon the in-built control system. This feature of work with new technology was the subject of considerable discussion in much earlier research on the question of automation. In a review of the literature up to the late 1960s, for example, an SSRC report stressed the importance of perceptual and conceptual skills, the latter being 'the most demanding and most critical for success' (1986: 16). Hirschhorn has argued strongly that such abilities become extremely important with modern technology: 'in cybernetic systems machines and workers complement each other with respect to a typology of errors: machines control expected or "first-order" errors, while workers control unanticipated or "second-order" errors' (1984: 72).

A third feature of the new technology was that it had to be programmed or instructed what to do. In three of our case-studies—brewing, chemicals, and insurance—programs rarely required modification and, on the few occasions that they did, the work was not done by the work groups under investigation. In contrast to the essentially routine nature of work in these cases, the wide variety of parts manufactured in our engineering plant meant that CNC machines required constant reprogramming, and this was done by operators. In this case, in addition to the 'second-order' control required when the machine was running (on this in the case of CNC machines, see Hirschhorn 1984), the worker still had a very significant degree of 'first-order' control over the production process, albeit mediated by the machine itself. As we have seen, the complexity of such programming varied considerably.

While the on-line processing system in the insurance company can be seen as assuming many of the functions previously undertaken by the underwriters, and while it did not require programming by them, they were nevertheless required to input certain data in order for the computer to undertake its work. Indeed, this was true in the other case-studies, for even in brewing and chemicals operators had to initiate stages of the production cycle. That is, the new technology did not operate a self-contained system of controlling the production process. The worker therefore had some degree of control over the production process. But in insurance the on-line system was significant in that, for the first time, it brought underwriters into contact with capital equipment more significant than pens and typewriters. In contrast, in the manual occupations studied it tended to distance the worker slightly from mechanical equipment.

While new technology changed the package of tasks which had to be performed if production was to proceed, the precise combination of tasks to make up a job was of considerable significance for work organization and experience. All of our case-studies highlight the significance of this point. Hence a comparison of brewing and chemicals indicates the way in which new technology of a broadly similar nature, although introduced into comparable production processes, was associated with very different changes in the occupational hierarchy. In chemicals it led to a polarization of job requirements and worker control, while in brewing it led to single-grade, largely unsupervised working. Similarly, if, as was possible at one point in the engineering company, more complex programming had been undertaken by planners rather than operators, the implications for the work experience of the latter would have been very different. Finally, we can contrast the decentralized system of on-line processing in the insurance company with the very different system employed in the subsidiary.

The significance of 'technical control' over workers was fairly limited. In none of our case-studies was it very great before the introduction of new technology, and it increased in less than half the cases distinguished in Table 3. Furthermore, many of the changes in controls and stimuli shown in Table 3 were not directly attributable to new technology. The higher quality standards in the automated chemical plant reflected the greater complexity of the chemical process—the reason for automation rather than its result. The same is true of the lager operators as compared with those in the old brewery. For the typical operator in chemicals, the operating rules and the system of supervision were more significant forms of immediate control than the technology. In the new brewery, the primary form of control was the work group itself, so that supervision was of much less significance than in the old plant. In this instance, it is true, technical control was among the more important constraints upon the worker, but this reflected the weakness of other constraints rather than the intensity of technical pressures. In our engineering study, the key form of control remained the quality inspector, but there were significant changes. In this instance, the pressures associated with piecework were removed and hence, in certain respects, management control became more important. However, despite the latter, we have rated supervisory control as less significant in the case of CNC than conventional operators. This is because supervisors lacked the knowledge and expertise to give CNC operators the technical guidance and direction they gave to those working on conventional machines. Finally, in insurance higher quality and output standards reflected the more general drive of management to rationalize policies, cut administrative costs, and reduce underwriting discretion, rather than being primarily attributable to the new technology. In the main, therefore, changes associated with the new technology were due less to the new technology than to the precise way in which work was organized around it, the fact that it did not impinge upon all the tasks undertaken, and the more general changes associated with technological innovation.

One other point concerning changing structures of control is also worthy of particular mention: the role of the supervisor. Except in the case of insurance, supervisors played a less important role with new technology. The engineering case has already been touched upon in the preceding paragraph. In chemicals, many of the supervisor's traditional functions were now assumed by the control room operator, who had a monopoly of the knowledge required for deciding upon the allocation of labour. In brewing, single-grade working not only reduced supervisory distinctions among workers but also made the group largely self-supervising.

More general changes, less intimately tied to task performance, also occurred with the introduction of new technology in our case-studies. In chemicals the extension of the job ladder meant fewer medium-grade jobs, but extended the occupational hierarchy still further by the introduction of the job of control room operator (and assistant control room operator). This might be seen, following Braverman (1974), as an example of management monopolizing technical skills; but it can equally be seen as increasing the promotion opportunities and hence life chances of manual workers. The changes in the occupational structure had implications for the structure of pay. Average earnings on automated plants were lower than on conventional plants since there were fewer middle-grade posts, but there were some opportunities for even higher pay (and improved fringe benefits) in the form of the post of control room operator. In brewing, single-grade working meant the disappearance of such intermediate posts as junior chargehand but greater opportunities of promotion to the position of chargehand. However, while basic rates of pay were the same for chargehands on the new and old plant, actual earnings on the former were typically lower, since there was no three-shift working and shift premiums were consequently lower. The pattern of shift-working reflected the level of demand relative to capacity rather than the nature of the technology employed. In our engineering plant, there was no change in the formal occupational structure or indeed in typical earnings. But there was an important modification to the payment system, since CNC operators were paid average earnings rather than having their pay based upon their own rate of output. This decision was in large part shaped by management's unwillingness to incur the costs of rating jobs on CNC machines. Finally, in insurance we find that gradings and pay increased and, consequently, so did promotion opportunities. In this instance, the change in the division of labour—underwriters being required to deal with all kinds of personal insurance in order to operate the new system efficiently—was formally deemed by management to justify the upgradings, but of possibly greater significance was the desire to win staff and union co-operation and commitment.

The introduction of new technology clearly led to greater productivity, or lower labour requirements per unit of output, in three of our case-studies, although in engineering and insurance these gains were significantly below those initially envisaged. In the brewery, gains in productivity were probably also achieved, although this is not clear from simple measures of manning relative to output, since the production process was considerably more complex than for traditional ales. However, only in insurance was new technology directly associated with job loss within the company, although in engineering a reduction in subcontracting led to some job loss in other companies. In both brewing and chemicals the new plants were additions to, rather than substitutes for, existing production facilities. But it should equally be noted that in three of our four case-studies overall employment fell during the period of technological innovation at the sites studied, and this was particularly marked in the cases of brewing and insurance. Here job loss was due to those pressures deriving from the product market which also led to the adoption of new technology. In the fourth case-study—engineering—while there was no job loss at the plant studied, there were substantial reductions in employment in the company more generally during the period of our research.

There were few other changes in the general pattern of labour regulation. In the engineering company it was formally agreed that traditional practices would be extended into the new issues associated with CNCs, and there was a marginal increase in the right to information which the unions enjoyed. In the longer term, there was the probability that more general changes might occur as a result of the slate of demands which the union had put forward. In the chemical and brewing cases, there were very few changes of any note. The exception was the insurance company, where the wide-ranging rationalization plans of management led to a spate of changes in the pattern of joint regulation. We shall consider this more fully below.

In overall terms, then, our findings indicate that new technology assumed certain tasks previously undertaken by workers, although it also required that the latter undertake certain new tasks. However, the precise impact of new technology depended to a significant degree upon how tasks were combined into jobs. Hence, similar innovations in essentially comparable basic production processes could lead to very different changes in work organization. Furthermore, there were many aspects of work upon which the new technology had little impact. Such changes as occurred in the broader work context—often indicative of important aspects of workers' life changes—were attributable to the organizational decisions made and to more general factors, in particular management reactions to market forces.

Our findings therefore suggest that while technology plays some role, other factors are of greater significance in explaining changes in work organization and labour regulation associated with technological innovation. It follows that theories of technological determinism receive very limited support from our findings. Furthermore, arguments concerning general trends in skills—whether in terms of increases or decreases—associated with new technology receive little support from our findings and have been shown to be too simple, so that they serve to conceal more than they reveal.

CHANGE AND CONTINUITY: THE QUESTION OF MANAGEMENT STRATEGY

In Chapter 1 we cast doubt upon the extent to which technological innovation could be understood in terms of a drive to control and subordinate labour. This criticism is consistent with our more general contention that employer priorities, even when they enjoy a degree of product market monopoly, extend far beyond problems of labour regulation and cannot be simply subsumed under such concerns. In particular, and important though internal contingencies may be, the major stimulus to action on the part of the private-sector employer is the product market. Signals from this market may highlight internal weaknesses and act as a stimulus to internal change. Hence, in all four of our case-studies, the state of the companies' markets was an important factor in the decision to introduce new technology. (As we shall see below, market considerations were also important in shaping the way in which new technology was introduced.) In the case of the insurance company, intensified competition in all areas and the increase in the attraction of personal lines relative to commercial insurance were central stimuli to a variety of policies adopted by management in which on-line processing was involved in an important way. A similar shift in market emphasis was found in the brewing company. Here traditional markets were declining while the lager market was expanding, and so the company decided to enter the latter: to do so required the construction of a new plant. The same was also found to be true in the engineering company. The decline in sales of new machines made spares production all the more important. Rapid provision of spares, an important element of after-sales service, could help the company to increase its share in the market, and, at the same time, there was a need for the company to ensure that it, rather than other manufacturers, provided them. Finally, and consistent with a long-standing practice, in the chemical company dedicated plants were constructed to manufacture products once the evidence indicated a sufficiently large market.

A second set of factors shaping management decisions to introduce new technology was of a largely technical nature. At one level there were at least some managers in each of our companies who were keen that their company have experience with new technology. In this sense, every company was dependent upon the general pattern of technological development. But the extent to which technological considerations entered into management thinking varied considerably. In the brewing company we have seen that management was keen to avoid the most advanced equipment for fear that it was not well tried and tested; consequent problems would endanger its primary goal of rapid entry into the lager market. The other companies showed rather more technical sophistication. In the insurance company there was a growing body of personnel with computing expertise who were keen to expand the use of new technology within the company and developed the software required for on-line processing (cf. Pettigrew 1973). Technical specialists in the chemical company had long been developing their own solutions to various problems. In the case of the automated systems of production, they had been trying to find technical solutions to yield difficulties for some time and had finally solved the problem. The construction of new plant provided an opportunity to apply these technical solutions. The engineering company also had a history of technological innovation, although in the case of CNC machines they were not developing their own technical solutions. In this instance, technical considerations included the fact that it was difficult to produce on conventional machines components which had initially been made on CNC machines in the company's main production plant.

In no case was the desire to control labour central to

management thinking. But the sorts of priorities which management had influenced labour regulation very directly in some cases. This is most clearly so in chemicals, where management had tried a variety of means to improve quality and reduce wastage. Tighter operating procedures and closer supervision, two of the techniques employed, were directly aimed at imposing greater control upon the working methods of operators. Automation in effect was a quite different strategy—it implied removing workers from key aspects of controlling the production process. The rationale—if not the reality—of automation in this case can be seen as being in many respects consistent with the argument that management was seeking to escape its dependence upon labour. However, it should also be noted that it was a concern over quality, rather than a fear of reliance upon workers directly, which was the key factor. Furthermore, we have seen that management in fact became more dependent upon a small number of key operators.

In the insurance company a central concern of management was the reduction of administrative costs. Given the importance of labour in these costs, there were very clear implications for the system of labour regulation. Management hoped to achieve dramatic increases in productivity, and, even with significant increases in the amount of business, initially envisaged substantial cuts in staffing. Furthermore, the related moves to tighten up underwriting rules had very clear implications for the degree of worker discretion. However, even in this instance it is important to remember that management approached the problem in systematic terms— they conceived the issue in terms of a market strategy and an overall review of the system by which insurances were processed. The substitution of labour and reduced worker autonomy were in many respects intimately bound up with this broader strategy and were clearly implied by it—but it was not the sole criterion guiding management's decisions.

In our other two case-studies questions of labour regulation figured less centrally in management thinking. It might, however, be suggested that the brewery company accepted certain labour control strategies which were implicit in the design of equipment, which was advertised as substituting automatic for worker control. But there is little evidence that this was an important criterion in management thinking: indeed, to the extent that they rejected the newest forms of equipment they also rejected the centrality of such criteria. The choice of CNC machines in our engineering company might equally be seen as accepting labour control criteria embodied in manufacturers' designs. This may be true in respect of certain tasks undertaken by the machine, but it should equally be remembered that a key factor was who programmed the new equipment and that this was not predetermined by the design.

More generally, what is striking about our findings is the limited extent to which labour considerations explicitly figured in management thinking, particularly in the earlier planning stages. This is consistent with the widely noted fact that, while there may be major implications for labour in management decisions, they frequently remain implicit and fail to be thought about coherently and in detail until fairly late in the day (see e.g. Batstone 1979; Batstone 1984; Winkler 1974). Moreover, in some cases the primary labour concerns of management involved not the details of work organization but broader questions of industrial relations. This was most strikingly so in the brewery company. Management in this case was concerned to ensure that any industrial relations problems among the contractors building the lager plant should not spread to its own employees and halt production in the old brewery. In addition, it used the new plant as a bargaining counter in its attempts to win changes in working practices in the old plant. In the insurance company the industrial relations implications of on-line processing were similarly central to management thinking. Its application was delayed on occasion for fear of worker and union opposition; the decision to upgrade staff working on the new system was also influenced by similar considerations.

As a consequence of the limited attention paid to the labour aspects of new technology, final decisions on work organization were often made rather late in the day, even after the plant had been installed. Hence, in the engineering company the final decision in favour of full operator programming was taken only several years after the first machines had been introduced. In the chemical company, the division of labour and the grading of jobs were varied on a number of occasions after the automated plant went on-stream. In the brewery the final manning pattern was decided only when construction of the lager plant was largely completed. In the insurance company there were changes on several occasions after on-line processing had been introduced. More generally, the case-studies show clearly that the reorganization of work is a continuing process, regardless of the technology employed; in other words, patterns of working introduced with new technology are not thereafter immutable.

Because of the lack of detailed planning early on, chance factors could play an important role in the final pattern of work organization. This is seen most clearly in the brewery study. Here, chance factors of importance included the realization that there would be spare computing power and, even more importantly, the delays in commissioning which led to management's realization of the advantages of single-grade working.

Given that management paid little attention to the labour aspects of technological innovation, the obvious question concerns the criteria which implicitly guided management decisions. The first point to note is that where the most dramatic changes occurred—the brewery company—they were not caused by an espousal of labour–management philosophies of a neo-human relations nature. It was in the engineering company that ideas of winning worker commitment to management ends were most clearly pursued, e.g. through experiments with quality circles. But such considerations did not, at least initially, inform management thinking on programming. However, the rationale which led to these experiments—buttressing the system of craft administration— did finally lead to programming being undertaken on the shop-floor. For while it would be misleading to suggest that it was carried out in these terms, underlying the debate on programming was the question of whether management could best achieve its ends through a reaffirmation or weakening of the system of craft administration (Stinchcombe 1970). The latter would have been the consequence of planners undertaking programming. We would suggest that it was far from coincidental that management finally rejected a more hierarchical control structure. First, the bureaucratic structure involved in the allocation and processing of jobs was less than effective. Second, the complexity and variety of work meant that management was very dependent upon the skills of its labour force. Any challenge to the autonomy of the craftsmen it employed could lead to serious consequences, not only in terms of industrial disputes but much more importantly in more subtle forms of withdrawal of goodwill. In effect the new technology was not such that a challenge to the craft ethos seemed, finally, to be a credible strategy. Thus, decisions in relation to new technology effectively served to confirm and strengthen traditional modes of regulation.

The same was true in the chemical case-study, where the principles of detailed division of labour and a finely graded occupational hierarchy were applied and extended. No thought was given to more radical changes in the organization of work. In the brewery, as we have just noted, more dramatic changes occurred, although management had initially started out with the intention of applying conventional working arrangements. In the insurance case-study there were significant changes in that the underwriting services section was merged with underwriting, and many staff were upgraded. The former change flowed relatively simply from the reduction in paperwork at area offices and the latter was shaped to a great extent by industrial relations considerations. But what is equally worthy of note is that management did not, as part of the move to on-line processing, seriously consider more dramatic changes in work organization (cf. Storey 1984; Rajan 1984). One such change—the move towards centralized processing—was being embarked upon towards the end of our research, but even this possibility had received scarcely any serious consideration at the design stage of the new system.

There are a number of factors which explain the degree of conservatism in management thinking. One, of course, is simply the limited attention paid to labour issues. The consequence of this is that traditional patterns were assumed to apply unless there were very clear pressures to act in any different manner (hence, for example, programming resulted in new principles in the engineering company). In this respect, as suggested in Chapter 1, existing practices had come to be largely taken for granted by management. In addition, management thinking was constrained by its structure and the characteristics of individual managers. There existed a balance of power and influence among management groups and, in the main, this does not appear to have been dramatically transformed by the factors leading to the decision to introduce new technology. In two of our case-studies there was no significant turnover of managers, so that the ideas and skills brought to bear upon the question of new technology did not change dramatically. In addition, even where outside specialists were brought in their influence was constrained by the parameters of management's initial decisions. A marked exception to this pattern, however, is found in the case of the insurance company. In this case, earlier computerization and associated developments had led to the establishment and subsequent expansion of a number of specialist functions. In other words, the structures of power and influence within the company were gradually changing. The appointment of a manager with a computing background to a post directly responsible for personal insurance then became the crucial catalyst for a change in the approach of the company. Changes in management also played some role in the engineering case-study.

A second factor constraining the degreee of change concerned the expectations and attitudes of workers and the unions. Hence, we have seen that in several cases industrial relations considerations significantly shaped management's approach. Changes need either to be acceptable to employees or to be compensated for in some way if confrontation or a lack of co-operation is to be avoided. Moreover, particularly when union organization is strong, existing agreements and understandings are likely to constrain management's scope for change. These might, of course, be renegotiated, but such a course of action is likely to be a long and costly exercise and might endanger the short-term success of technical innovation.

As was noted in Chapter 1, some of these constraints could be overcome. Most obviously, on green-field sites management are likely to have much greater freedom of action. Such options are often far more difficult to pursue on existing sites. Changes may in some cases have to be applied 'across the board', or else the new technology has to be isolated in some way. But even in the latter case—which was true in part in both our chemical and brewing studies and was also an option in the engineering company—it is necessary to prevent the application of traditional agreements and understandings. This is likely to meet with opposition, particularly where new arrangements are less favourable to workers. A small number of employers have, nevertheless, for a variety of reasons, been prepared to risk confrontation in the name of change. In the main, however, changes appear to have been relatively marginal (see e.g. Batstone 1984; Batstone and Gourlay 1986). All four of our employers considered themselves, with some justification, to be fair and reasonable employers who wished to carry their employees and unions with them in any change.

A further constraint in our case-studies was that the new technology was, in every case, manned by existing staff. This was not inevitable, and in principle it would have been possible to recruit new staff. But where the new technology was replacing existing equipment this would have raised major problems concerning redundancies. In a number of cases management did seek scope to select staff in new ways: but in the brewery outside recruitment was rejected by the union, and in insurance the union opposed the treatment of upgraded underwriting positions as new jobs. In these cases, management was going against the traditional understandings and agreements embodied in the internal labour market. Such action therefore raised more general issues and managements deemed it sensible to avoid such breaches in the pattern of mutual accommodation.

A final factor explaining the limited degree of change is that the basic characteristics of corporate activity remained constant. While, as we have noted above, several of our case-study firms were planning to enter new markets or to shift the balance of their efforts within particular markets, these changes were relatively marginal. The nature of the new market emphases rarely involved any fundamental changes in the broad approach of the company: in market terms lager was not so different to the traditional beers produced by our brewery; the insurance company had for many years dealt with personal insurance; the products in the automated chemical plant were aimed at the same sorts of markets as the company's older products; and the engineering company had always produced its own spares. Markets were changing in terms of the signals they transmitted to the companies, but their broad characteristics remained the same. While there was need for adjustments, changes in the basic skills required of the labour force were of a fairly limited nature. Certainly some changes were required, e.g. an ability to program in the engineering company. Nevertheless, even in this instance the traditional skills of the setter-operator were still of central importance, and it was certainly easier to learn to program than to learn these traditional craft competences. In these respects, to the extent that previous patterns of labour regulation had been deemed adequate they were, in broad terms, likely to be considered equally relevant and useful in the new situation. Thus, added to the pressures internal to the company which favoured the limitation of change were important and constant features of the market and the sorts of skills which were required to meet its demands.

There were, then, a series of factors which encouraged management to minimize the scale of change in labour regulation when new technology was introduced. However, it should be noted that this does not mean that managements do not change their approach to labour regulation. Indeed, in two of the cases studied there were much broader changes occurring at the time that the new technology was introduced. In the insurance company we have noted that the move to on-line processing was merely one of a series of strategies being pursued by the company which had important implications for labour regulation. In the brewery there were also major changes being negotiated in working practices. In both cases, at least as far as industrial relations considerations were concerned, new technology was not the central concern of management. As a consequence, it seems that management sought to ensure that issues arising in relation to new technology did not exacerbate the general situation. They were deliberately kept 'low-key', although matters inevitably became intertwined. Job loss, for example, arose over on-line processing and other management initiatives in the insurance company.

In summary, the way in which new technology was introduced was shaped by the basic nature of the markets in which the company operated, the internal structures of management and the labour force, and the patterns of joint regulation. These elements are closely interrelated and typically serve to act as constraints upon each other. Thus, not only may there exist strong pressures towards conservatism from within the company, but these may be reinforced to the extent that key conditions of market survival remain constant. In this respect, our case-studies provide broad support for the arguments we developed in Chapter 1 concerning the factors which are likely to shape management's approach to change and to labour regulation.

UNION INFLUENCE AND STRUCTURE

The preceding chapters have devoted considerable attention to the degree of union influence over the introduction of new technology. Four findings are of particular importance. First, in overall terms the range of union influence was limited. Although the unions did not play a significant role in shaping general management strategy in any of the case-studies, they were informed, and even consulted, about the decision of management to introduce new technology. Some workers or stewards frequently played a small role in the selection of equipment or aspects of design. However, the union role was essentially confined to matters of immediate relevance to the wage–effort bargain. Second, the degree of union influence varied significantly between the various case-studies. In the chemical case-study, for example, the role of the union was negligible, while in the engineering and brewing case-studies many aspects of work organization were jointly agreed. Third, with the partial exception of BIFU in the insurance company, the pattern of union influence in relation to new technology was broadly comparable with its more general role within the workplace. Fourth, the degree of union influence was closely related to the nature of union organization.

In *Unions, Unemployment and Innovation* it was argued, on the basis of survey evidence, that there was a close link between the range of bargaining over new technology and that over more general issues (Batstone and Gourlay 1986). The primary emphasis in that study was placed upon union organization and strength. Basically, it was argued that if unions were strong enough to bargain over e.g. manning arrangements then, *ceteris paribus*, they would be able to do the same in new technology areas. Our case-studies broadly support this view, but they also indicate a number of other important factors which could not be gleaned from the survey data. First, as we have noted above, existing agreements often cover many questions relating to new technology. Thus, unless management is seeking to use new technology as an opportunity to transform the pattern of joint regulation, some issues are not likely to require any significant debate. The assumption is likely to be that normal practice will apply. If changes are mooted—as they were in some instances in our case-studies—they may not be seriously pursued in the event of opposition, since they would have implications not only for the specific situation under consideration but also for more general practice. They may therefore be seriously contested, not simply because they may be seen as the 'thin end of the wedge', but also because they may be defined as a sign of 'bad faith' on the part of management. What might be relatively minor issues in themselves may therefore assume much greater significance and adversely affect the general climate of industrial relations.

The tendency towards continuity in forms of labour regulation helps to explain a further striking feature of some of our case-studies. This is the lack of trauma associated with the introduction of new technology. In the chemical case-study this was also due to the small numbers involved and limited union influence. In the brewery, and to a lesser extent in the engineering company, the process of bargaining over many issues seemed remarkably straightforward and simple. In part this reflected the skills of the negotiators who understood the micro-politics of their counterparts, but it also reflected the fact that a great deal of negotiation was really little more than running through a check-list of issues, ensuring either that traditional rules would apply or that the proposals of one side were readily acceptable to the other. This is not, of course, to suggest that there were no issues of contention. Indeed, in the engineering company there were a number of very significant areas of disagreement, e.g. operator programming, the new technology agreement, and the broader demands of the union. Moreover, sanctions were threatened or employed in relation to the first two of these. In this case, however, there were two key factors. First, the question of programming raised a new principle which might seriously have endangered the traditional autonomy of the craftsman. Second and relatedly, the union saw the introduction of CNC machines as an opportunity to win commitments from management on a variety of other issues. Such strategic thinking in part reflected the significance of the new technology and in part the sophistication of union organization.

In the insurance company industrial relations issues were rather more contentious. The widespread changes which the company was seeking to introduce raised a whole series of questions which had never before been tackled by the company, many of which appeared to endanger the implicit contract between employer and employee concerning a secure career structure. However, it is open to question how far these issues would have surfaced had not significant changes in union structure been occurring at the same time.

A second factor which helps to explain the relative stability of the pattern of joint regulation has also been touched upon in the previous section. This concerns the way in which particular 'ways of doing things' come to be part of managers' assumptions. This is equally true of shop stewards and workers. Furthermore, in many situations—as in our brewing and engineering studies—it is clear that these assumptions are to a significant extent shared by the key negotiators. This is the essence of a strong bargaining relationship, permitting the 'sounding out' of options and facilitating the understanding of each party. Not only are there prerequisites, in terms of the significance and influence of each party to the relationship, but it is also true that such relationships take time to develop.

Each has to get to know the other's personality, and it is only over an extended period, and after a wide range of bargaining, that some degree of consensus builds up as to the balance of power and the stratagems which are likely to be effective. In this way, therefore, a degree of stability tends to be fostered by the very existence of shared understandings and strong bargaining relationships.

Such relationships may frequently be important because they constitute a means of introducing a degree of certainty and predictability into industrial relations. This is to be seen most clearly, perhaps, in our insurance company. There the strong bargaining relationship between the division secretary and the personnel manager provided a means of dealing with the uncertainty which both shared concerning the possible reactions of a membership which was characterized by divided loyalties and an ambivalence concerning collective means of interest representation and pursuit. Such uncertainty was further increased by the fact that management plans raised new issues which had not been encountered before in the company.

Strong bargaining relationships also help to explain why disagreements often took so long to resolve. Such delays were in part attributable to other problems involved in introducing the new equipment. This is a common problem with new technology (see e.g. Rhodes and Wield 1985). Delays and unforeseen circumstances frequently changed the terms of the discussion of contentious issues or led to changes in the views of the parties concerned. But, even allowing for these, our case-studies indicate numerous occasions when the resolution of relatively important issues took a long time. What is interesting here is that a good deal of the industrial relations literature stresses the need for rapid procedures to resolve disagreements. This may be true in many circumstances, but there are other occasions when delays can be conducive to resolution. This is so for a number of reasons. First, given that views change and that the balance of power shifts, one party may gain advantages by waiting until the other side appears to be more receptive to its proposals (see e.g. Batstone *et al* 1977). Second, delays may provide an opportunity to float suggestions and ideas informally and often indirectly. In this way, each party may develop a more sensitive understanding of the parameters of the other's position, thereby facilitating gradual movements towards agreement. Third and perhaps most importantly, each party may delay pushing for a resolution in order to permit intra-organizational bargaining to occur (e.g. Walton and McKersie 1965). This was important in both our engineering and brewing case-studies. Ideas and opportunities would be floated and 'nudged' by key union officials in order to shape the process of negotiation within management. Fourth, in the event of disagreement delays are an alternative to manifest conflict. They indicate an uncertainty over the outcome of conflict and/or hesitancy over the extent to which resource mobilization is feasible or justifiable in relation to the issue. For example, in the engineering case-study management at one stage declared its intention to issue a unilateral statement on the question of new technology. However, the threat of strike action led them to retreat from this plan of action.

Even the case of the insurance company is consistent with our argument. For while it is true that the range of union bargaining increased with the introduction of new technology, that change was not attibutable solely or even primarily to the new technology. Rather it reflected two factors. The first of these was the range of issues and problems arising out of other changes associated with management attempts to improve competitiveness. Hence the correspondence between the bargaining over new technology and the more general pattern of bargaining. Second, over the period studied there had been significant changes in union organization. Membership was transferred from an in-house staff association to BIFU; union density rose and a variety of organizational innovations were introduced. The key stage in this pattern of change was joining BIFU, and this occurred largely independently of the rationalization initiatives of management. That is not to say, however, that the uncertainties which the latter created for staff were unimportant in the development of the role of the union. But it is open to serious doubt how far the old staff association would have achieved the same degree of influence in similar circumstances. Hence, the insurance case-study supports the argument that a relationship exists between bargaining patterns and union sophistication.

The changes which occurred in the role of the union in the insurance company do, however, highlight another theme which received insufficient emphasis in *Unions, Unemployment and Innovation*. The survey reported in that study did not permit any analysis of member attitudes or, indeed, of detailed aspects of the work situation. However, what is striking in our case-studies, and particularly in the changes occurring in the insurance study, is the way in which the interrelationship between union organization and union influence is mediated by the pattern of day-to-day work experience and its influence upon member attitudes. This theme can be usefully developed by briefly considering the debates on factors leading to the growth of white-collar unionism.

Great emphasis has been placed by many sociologists upon the proletarianization of white-collar work as an explanation of the growth of white-collar unionism (e.g. Crompton 1976 and, specifically in relation to insurance, Crompton 1979). The introduction of new technology is often seen to be an important factor in this process (e.g. Glenn and Feldberg 1979; de Kadt 1979). However, such arguments encounter a number of problems concerning the timing of such presumed proletarianization on the one hand and unionization on the other (e.g. Heritage 1980). Our findings suggest, for example, that many features of such proletarianization as had occurred date back many years, and that the extent to which new technology was a cause of further movements in this direction was limited. Moreover, while a sense of insecurity and uncertainty undoubtedly developed among many staff in the face of the changes occurring in the company (and there were some elements which might be seen as constituting a movement towards proletarianization), other elements of their work and market situations shifted in the reverse direction. The average salaries and other conditions of service of underwriters also remained substantially above those of manual workers. Our case-study suggests that other factors are important in explaining the growth of union organization. As many industrial relations commentators have noted, while factors relating to proletarianization may be important, so also are the strategies pursued by key actors—the state, the employers, and the unions themselves (e.g. Bain 1970). It should be noted that much of the increase in union density in our insurance company occurred when government sympathy for trade unionism was notable only for its absence. The employer was prepared, however, to recognize the union and adjusted its approach to industrial relations to some degree. But what was central was the transfer of membership to BIFU, and we have seen that this was due primarily to chance factors—the company taking over a firm where a union was already recognized—and problems of administrative overload (cf Adams 1975; Price 1983). The move towards greater 'unionateness', at least in organizational terms, therefore derived to a significant extent from essentially organizational concerns rather than dramatic shifts in the attitudes of the typical employee. Furthermore, we have suggested, it was the growth of organizational sophistication which, along with the growing sense of uncertainty among staff, led to the increase in union density. The final strand we would wish to emphasize in this context, however, is the way in which the growth of membership was related not only to recruiting drives etc., but also to the way in which the union was gradually able to show that it could affect members' interests. In other words, not only were the work and market situations of the staff undergoing certain changes, but they were also becoming increasingly subjected to union influence. In this way, the day-to-day experience of work was increasingly of a nature to encourage awareness of, and resort to, collective representation. Various aspects of the work situation were now subject to rules, regulations, and agreements which were jointly made by management and unions.

In our insurance company, of course, this process was only beginning. But the study does highlight the way in which there is a close interaction between membership levels and attitudes, union organization and strategy, and the basic situation in which employees find themselves. It is this ability to achieve a certain 'mobilization of bias' (Schattsneider 1960) in favour of collective representation which provides the unions with a firmer base among the membership. In turn, member commitment and organizational sophistication help

TABLE 4. Union influence over patterns of labour regulation compared with union sophistication

	Chemicals	Brewing	Engineering	Insurance
1. *Influence over constraints on task performance:*				
Job design/content	—	√	√	—
Output standards	—	—	—	—
Quality standards	—	—	—	—
Operating rules	—	√	√	(√)
Discipline and supervisory action	(√)	√	(√)	—
Technical control/stimuli	—	—	√	(√)
Financial incentives	n/a	n/a	√	—
NO. OF AREAS OF INFLUENCE	.5	3	4.5	1
2. *Influence over labour usage:*				
Recruitment	—	(√)	(√)	—
Promotion (in group)	(√)	(√)	n/a	(√)
Transfers	(√)	(√)	(√)	—
Flexibility	(√)	√	√	—
Use of 'secondary' labour	(√)	√	√	(√)
Manning levels	—	√	√	—
Training/experience acquisition	—	√	√	—
Redundancy	—	√	√	—
Overtime/reliefs	—	—	√	—
NO. OF AREAS OF INFLUENCE	2	7.5	7	1

3. Influence over worker rewards and conditions:

Health and safety conditions, etc.	(√)	√	√	√
Hours of work	√	√	√	√
Payment system	√	√	√	√
Pay levels	√	√	√	√
Grading, job structure	√	√	√	√
Fringe benefits	√	√	√	√
Disciplinary system	√	√	√	√
NO. OF AREAS OF INFLUENCE	6.5	7	7	7
4. Influence over broader management strategy:	(—)	(√)	(√)	(√)
Union sophistication:				
100 per cent union membership	√	√	√	—
Representative steward constituencies	√	√	√	√
All steward positions filled	—	√	√	—
Regular meetings of stewards	—	√	√	—
Senior stewards	√	√	√	√ (company)
Full-time stewards	(√)	√	√	√ (company)
SCORE	3.5	6	6	3

Note: (√) counted as .5 for number of areas of influence.

to extend and confirm the range of union influence and control. We have seen that incomplete membership and member ambivalence—reflecting the nature of the work situation—limited the degree of organizational sophistication. Representative positions remained unfilled, and local structures of co-ordination remained weak. This meant that the extent to which the unions were able to impinge upon the day-to-day experience of members, particularly on matters of work organization, remained limited.

This 'mobilization of bias' is, we would suggest, important more generally. Batstone *et al.* (1977), for example, pointed to its significance, as did Edwards and Scullion (1982), albeit in somewhat different terms. Our other case-studies also suggest its importance. Table 4 looks at the areas of union influence. It can be seen that in the brewing and engineering case-studies the unions had a considerably wider, and generally deeper, influence over factors relating to task performance and the pattern of labour usage. Variations were less marked concerning worker rewards and conditions. However, it is the former which are most likely to be significant in encouraging the worker to think in collective terms. Hence, while there was 100 per cent union membership among the manual groups studied, we have noted that in the chemical case-study many workers defined the union as having little role in matters concerning work organization, other than as a source of information to guide individualistic strategies. In both engineering and brewing, workers were likely to be much more aware of the role which collective representation played. Such awareness supported and reflected a high degree of organizational sophistication on the part of the workplace organization. All steward positions were filled, and union organization was co-ordinated and united in a variety of ways. In contrast, in the chemical company—as in insurance—there were unfilled steward positions and weaknesses in the structure of co-ordination. As a result, there were fewer pressures to develop workplace rules which could buttress the role of the union at shop-floor level, and little support for stewards who attempted to extend their influence locally.

Thus the extent to which the work situation reflects

individualistic rather than collective factors appears to be an important concomitant of union sophistication. But this is not to say that strong union organization demands that individualistic worker strategies be eradicated. It is rather a question of balance and the extent to which the union shapes individualistic opportunities and strategies. Our case-studies can be used to illustrate this point. In all of the companies studied some form of internal labour market existed, in the sense that higher positions were normally filled by internal appointments rather than from the wider labour market. The exact significance of such career structures varied widely—it was clearly most important in the insurance company and least in engineering. With the latter exception, all the companies employed a quite finely graded occupational hierarchy and hence a career route. Moreover, in all of the companies there was scope not only for vertical but also for horizontal mobility, which may or may not be associated with increased earnings (e.g. through shift premiums). The significance of the latter appears to have been largely ignored in the literature, but our evidence suggests it may often be a more significant factor than vertical mobility. It can provide an important means by which workers can overcome a wide variety of dissatisfactions which they might find in any particular job; this, we noted, was particularly true in the chemical company. More generally, there was plenty of scope in all four case-studies for workers to discuss and 'negotiate' issues individually with supervisors. Indeed, in the engineering company individual bargaining over a multiplicity of features of the piecework system was very marked (cf. Brown 1973). In brewing and in engineering, however, these individualistic options did not weaken the role of the union. Two reasons appeared to be important. First, the union clearly played a role in shaping these individualistic options. In the engineering plant, the union negotiated the parameters within which individual bargaining occurred and might also become involved in cases of disagreement. In the brewery, the union had negotiated over a variety of aspects of the transfer and promotion procedure. Second, the scale of joint regulation provided alternatives to individualism, so that the internal labour market and individualistic strategies assumed less significance in the brewery, for example, than it did in chemicals.

Two more general points are suggested by these findings. The first is that, as Batstone *et al.* have argued more generally (1984), the significance of what is normally termed bureaucratic control structures varies according to the extent to which they are influenced by trade unions. Internal labour markets are likely to be less inimicable to trade unionism where they are both shaped by the unions and where trade unions provide alternative routes for the promotion of self-interest. It is misleading, therefore, simply to note the existence of different forms of control. What is of crucial significance is not only the variety of forms of control, but also the degree of union and management influence within and over these different forms.

The second point which the preceding discussion raises concerns the extent to which there is an inherent conflict between individualism and collectivism. Certainly there are tensions between the two, but it can be suggested that the extent of any contradiction between them has often been exaggerated. For example, the growth of trade unionism in the insurance company was at least in part attributable to a desire on the part of workers to defend individualistic career routes. This theme can also be usefully pursued at a more general level.

It has long been commonplace for craft unions to permit individual negotiation over rates of pay even where no incentive scheme exists. Such unions have classically been concerned to ensure that a certain minimum rate is paid but have not set a maximum figure. In this sense, a merit component has often existed in the pay of craftsmen, which is comparable in some respects with that classically found among white-collar workers. Moreover, craft unions have often insisted that their rank-and-file members be supervised only by craftsmen, with the effect that some promotion opportunities exist. They have also been ready to accept employers, at least those working 'on the tools', into membership. On the other hand, many white-collar groups have traditionally stressed collective forms of regulation. This is most notably true of many professions, while technicians have often sought to mimic the strategies of their manual counterparts. Particularly in the public sector, other white-collar groups have often been concerned to protect and regulate career routes. They have sought to limit ports of entry, to lay down rules concerning qualifications for promotion, and to ensure that distinct hierarchies are retained. In many respects, therefore, their strategies can be seen as comparable to those of craft and other manual unions concerned with protecting a particular job territory (cf. Rubery 1978; Jacoby 1984; Elbaum 1984 on the role of unions and work groups in the establishment of internal labour markets). These comparisons suggest that, even if craft unions are on the decline, other forms of trade union may seek to establish broadly similar claims to territories and, along with this, to demonstrate a concern with training and the more detailed aspects of work organization.

But there are, equally, important differences between craft unions and white-collar unions. The former have classically sought to preserve job territories and control across a national labour market; they have been concerned with the interests of particular occupational groups within a wide range of companies. Often that group has constituted a small proportion of the total labour force. In contrast, most white-collar and many manual unions have focused their concerns with job territory upon the individual plant or employer. In other words, members' life chances are defined primarily in terms of an internal labour market, rather than in terms of a national labour market. However, particularly where the union represents a large proportion of workers in any one firm, and where promotion routes cover a wide range of occupations, this employer orientation may mean that the union seeks a wider range of influence over corporate affairs than is typical of craft unions. Under these conditions, internal labour markets and union sophistication and scope may combine to foster a significant degree of union influence and a broader perspective on the part of the union. This may lead to a significant degree of co-operation with management goals—as in our case-studies—but such co-operation is by no means guaranteed: it is conditional upon the gains which are achieved. Where these are deemed insufficient, the very factors which were once conducive to co-operation may become potent stimuli to major challenges to the employer. This, it would seem, is an important explanation for the growth of strike action among previously quiescent white-collar workers—strike action was supported by the membership in our insurance company for example (see also Batstone and Gourlay 1986).

The preceding argument is in large part conjectural, and there are serious questions concerning the extent to which white-collar unions will indeed become sophisticated and influential. But it does indicate the possibility that, even though the general nature of trade unionism may be changing in certain respects with the decline of areas of traditional union strength and the increasing importance of white-collar work, it is by no means self-evident that trade unionism in the longer term will become weaker. This connects with the argument, put forward in *Unions, Unemployment and Innovation*, concerning the potentially more sophisticated and integrated structures of white-collar unions (Batstone and Gourlay 1986).

The scope and nature of trade union organization and action, it is widely argued, are also strongly influenced by the strategy of the employer and the general market situation. It now seems generally accepted that few employers have sought to challenge the institutional position of the unions. Somewhat more common, however, have been attempts to bypass the unions in some way. In three of our case-studies we have noted such attempts, which in the main were successfully resisted by the unions and hence had little significance. It has, however, frequently been suggested that union influence has declined as employers have reduced the degree of joint regulation. Some care is required in putting forward such arguments. In the first place, as has been widely noted, it is notoriously difficult to measure power. A second problem can best be discussed by an analogy. If we know that a weight-lifter is able to lift fifty kilograms and then we discover that he cannot lift a hundred, we would not conclude that he had become weaker. It is likely that we would argue that he never had been able to lift such a weight, and that this became evident only when he was asked to do so—although we may have been convinced of this fact for a long time. However, there would be conditions under which we might say that this weight-lifter had become weaker. If, for example, the attempt to lift the greater weight had led to a serious injury with the consequence that the weight-lifter was subsequently no longer able to lift fifty kilograms, then few would dispute that he had become weaker. Similar arguments can be applied to trade unions. That is, it would be possible to argue that at any time in the past, if unemployment had reached several millions and if there had been legislation which impeded the scope of union action, unions would have been unable to resist these pressures. From this perspective, the question is whether or not such forces have served to reduce the ability of the unions in the present and future to handle the sorts of problems they did in the past with the same degree of effectiveness. This has, of course, been a topic of considerable debate. While there is some disagreement, it seems that trade unions, while weaker, have certainly not been decimated. Membership has declined, but not dramatically, as a proportion of employees actually in employment. The evidence indicates that trade union organization in the workplace has generally remained intact, although several unions have encountered serious financial problems nationally. While many trade union members are currently more ready to co-operate with management than they were in the past, it is far from evident that this perspective will be maintained in any upturn. Moreover, it should be remembered that such co-operation has long been a feature of many trade unions.

A third and related issue concerns the question of what influence trade unions have actually been able to wield in recent years. Following from the previous point, it is necessary here to make a distinction between union influence and worker interests. It is possible for union influence to remain constant or even to increase, at the same time as their members' terms and conditions deteriorate. This is perhaps seen most clearly in our insurance case-study. There, we have seen, the scope of union influence had increased significantly in recent years, even though certain aspects of the work and market situations of staff had deteriorated. The key question, of course, is what would have happened if there had been no union—would members' conditions have worsened still further? We suspect that the answer is that they would have.

In other cases we also see some deepening of union influence. This was so in the engineering case-study, where the union won a series of commitments from management on a range of issues, as well as securing the principle of operator programming. In our brewing case-study, the range of union influence increased only marginally. But, we noted, in the process of negotiating more general changes in working practices, the depth of union influence increased. Whereas negotiations had once been fairly centralized, multi-level bargaining now became more significant. Shop stewards negotiated more, and in a co-ordinated manner. Contrary to what is often argued, therefore, productivity bargaining was associated with a decentralization, not centralization, within the shop steward body (for a review of these arguments, see Batstone 1984). Moreover, shop stewards had maintained a higher level of bargaining activity on the shop-floor and had been able to persuade management to modify their proposals significantly. Crucial to both the deterioration in members' conditions and the expansion of the role of the unions was the fact that changes in the market situation had led to new issues being placed on the agenda by management. However, there is little evidence in our case-studies of union influence declining—if anything the reverse—and there is no evidence that the unions were being bypassed in a way which was distinctively different from what had always occurred.

One reason for this, of course, was that the unions in our case-studies did not present any fundamental challenge to the proposals for change put forward by management, although they contested the details—sometimes with considerable force. In several cases these proposals, unpalatable though they might be, were seen as necessary given the product market situation. In other words, it was believed that members' interests would best be served by accepting the basic approach of management. It is certainly possible to argue that stronger unions would have been able to put forward proposals which would have better served members' interests; this is, indeed, the case. But their failure to do this is a reflection of a long-standing weakness of most British trade unions and workplace organizations: it is not a new phenomenon. What is clear, however, is that where union organization centrally was more sophisticated—that is, with the exception of the chemicals case—the unions did look at and think about management proposals seriously. As a result, the engineering union, for example, sought a variety of guarantees, many of which it partially obtained. There may, however, have been another important factor—that the unions believed the balance of power was such that concerted opposition to management plans would have led to defeat.

In all of our case-studies, the unions' basic approach towards management was essentially a co-operative one. Strike action, for example, had been either unknown or rare in the past. The question therefore arises of the extent to which the unions were incorporated—imprisoned in a managerialist logic. Such a view might seem particularly plausible given the time and conditions under which trade union organization had developed. The engineering plant had been established in the early 1970s, and very early on, without any real pressure from the union, management had agreed to a structure of shop stewards. In the chemical company the union was recognized at about the same time; while membership had been growing, it is unlikely that shop steward organization would have developed to anywhere near the same extent had management been opposed to such developments. In the insurance company, while an important stimulus to the extension of union organization was the move to BIFU, management was again prepared to accept a more sophisticated form of union organization. Even in the brewery, where union organization had a much longer history, important changes in the pattern of shop steward organization occurred in the late 1960s.

In many respects (although to varying degrees) it might therefore be suggested that our case-studies provide classic examples of what have been termed management-sponsored shop steward organizations. These, it has been argued, became common in the 1970s as part of the general reform of workplace industrial relations, particularly in such sectors as chemicals and food and drink, and other areas where there was not a tradition of shop steward organization. Such organizations, the argument runs, appear to be comparable to more traditional forms of shop steward organization in formal terms, but act as a kind of control over, rather than for, the membership—that is, they are effectively an arm of the personnel department. Thus, not only do they adopt a co-operative approach towards management, but they also fail to exert any influence on work organization matters, confining their activities to the wage side of the wage–effort bargain (Hyman 1979; Terry 1978; Willman 1980).

Our engineering and brewery cases clearly contradict this argument. We have shown that the shop stewards played an important role in many aspects of work organization. It might be countered, however, that these are less than ideal examples of management sponsorship. The engineering industry, and indeed the larger company which built the plant studied, had a long tradition of shop steward organization. The brewery case might be seen as a less than ideal example simply because some form of steward organization had existed for many years. These arguments are only partly valid. It has been claimed, for example, that long-standing steward organizations might be co-opted by management (e.g. Willman 1980). In addition, green-field sites with largely 'green' labour, particularly in areas without a militant tradition (as was the case with the engineering company), might similarly be expected to provide an ideal opportunity for management sponsorship.

The insurance and chemical cases, however, do appear to fit the thesis much better, since the unions had little influence over work organization matters (although it was not totally absent). However, in these cases we would explain the limits to joint regulation in terms of the details of union structure, rather than in terms of the fact of management sponsorship. While in many respects union organization was sophisticated, there were marked weaknesses, and in both cases these stemmed in large part from member attitudes. In one sense it follows that steward organization would not have been as developed as it was if it had not been for management support. It is therefore quite correct to distinguish them from other steward organizations. However, as has been argued elsewhere (Batstone 1984), the crucial question is whether workers' interests would have been as effectively defended and promoted if there had been no such sponsorship. We are confident that they would not have been. In chemicals, for example, the union did play some role, and in addition its activities had led to substantial improvements in wages. Furthermore, we have seen that the insurance union was beginning to extend its range of influence. Our earlier discussion indicates that to focus upon the fact of management sponsorship runs the risk of trivializing the factors which explain the role which shop stewards play. What is more important, we have suggested, is the interaction between management strategy, union sophistication and strategy, and the nature of the work situation. In short, theses of management sponsorship contain weaknesses comparable with those found in arguments concerning the incorporative effects of the bureaucratization of shop steward organizations.

Certainly our case-studies provide little evidence of any weakening, either in the institutional position or the range and depth of influence of workplace trade unionism. Where changes had occurred, they were in the direction of greater sophistication and influence. With some notable and far from unimportant exceptions, this would appear from the available survey data (Batstone 1984; Batstone and Gourlay 1986) to be the more general pattern. Our case-studies provide support for the view that the range of influence of shop steward organizations is intimately related to their sophistication. Furthermore, as we argued in *Unions, Unemployment and Innovation* (Batstone and Gourlay 1986), there are strong relationships between the general pattern of bargaining, the range of bargaining over new technology, and the degree of union sophistication. However, what we have stressed more strongly here than in our previous study are two important elements—the mobilization of bias which sophisticated organizations can introduce into workplace arrangements, thereby strengthening member commitment, and the typical continuity of industrial relations which, unless strong pressures exist, will tend to spread from traditional to new areas of work within an establishment.

SOME COMMENTS ON PATTERNS OF LABOUR REGULATION

In the preceding sections of this chapter we have, at various points, touched upon problems concerning general approaches to the analysis of labour regulation or labour control. We do not intend to repeat them at length here, but it is useful briefly to recall them and to place them in a more general context.

In discussing the details of task organization, we noted that a variety of control forms existed in relation to particular occupational groups. As was noted in Chapter 1, this automatically raises questions concerning the utility of simple categorizations of the kind employed by Edwards. Hence, for example, to the extent that there is any pattern there is a tendency for the degree of supervision to be positively associated with the extent to which work methods are bureaucratically specified; Edwards's arguments would lead one to expect the reverse (1979). Similarly, there tends to be a positive relationship between technical and bureaucratic control at the task level, and between technical control and supervision. In other words, if one were forced to opt for some general statement concerning relationships between different forms of control, one might be inclined to see them as cumulative rather than alternative. A more satisfactory approach, however, would be to note that the significance of different forms of control and their combination may vary widely.

In Table 5 we look at aspects of the broader patterns of labour regulation which have been stressed in a good deal of recent writing—elements of bureaucratic control, as defined by Edwards (1979), and various other means which are seen as fostering worker commitment and loyalty. On the first theme, we can note that bureaucratic control tends to be greatest in the case of insurance and least in the case of engineering. The categorization is useful only in differentiating between the engineering company and the other cases. But, as we have noted, there are extremely important differences between, for example, the chemical and brewing companies which this classification totally fails to catch.

Similarly, the overall pattern of techniques aimed at fostering individual loyalty to the company varies only slightly between the firms. Furthermore, the precise groupings vary—for example, only the engineering company employed individualistic involvement measures, such as quality circles and briefing groups. Again, therefore, such measures do not

TABLE 5. Aspects of the broader structure of labour regulation

	Chemicals	Brewing	Engineering	Insurance
(a) Bureaucratic control:				
Finely graded hierarchy	√	√	—	√
Formal job descriptions	√	√	(√)	√
Formal rules for supervisors	√	√	√	√
Incremental structure	—	—	—	√
Formal grievance procedure	√	√	√	√
NO. OF FEATURES	4	4	2.5	5
(b) Fostering worker identity:				
Consultative structures	√	√	√	√
Communication systems	√	√	√	√
Individualistic involvement measures	—	—	√	—
Merit payments	—	—	—	√
Bonuses related to company performance/ share-schemes	√	√	—	√
Company-based social activities	√	√	√	√
NO. OF FEATURES	4	4	4	5

Note: (√) counted as .5 for number of areas of influence.

appear to differentiate satisfactorily between our companies.

There is a notable lack of correlation between these broader forms of labour regulation and task control features. There is no clear tendency, for example, for bureaucratic rules at the general level to be associated with their significance at the task level. Similarly, there is no clear relationship between a high level of general bureaucratic control and low levels of technical or supervisory control. Again, to the extent to which there is any pattern, it would appear to be a tendency towards the multiplication of different forms of control.

It is equally important, however, that these sorts of categorizations fail in the main to note what preceding chapters suggest to be crucial differences in forms of labour regulation. This is so for a number of reasons. First, they do not take account of the scale of joint regulation. As illustrated in Table 4, the areas of joint regulation serve to differentiate sharply between the brewing and engineering companies on the one hand, and the insurance and chemical companies on the other. Union influence can change the significance of rules and controls. Hence, for example, many of the rules in the insurance company find their origin in management initiatives. On the other hand, in the brewery many rules are, at least in part, the product of union initiatives, as was clearly indicated in our discussion of productivity bargaining. In the former case, the rules tend to be management tools, while in the latter they act much more as constraints upon management. We return, therefore, to a theme stressed in the preceding section; namely, the way in which one form of control can transform the significance of another. In this instance, the degree of joint regulation changes the significance of bureaucratic forms of regulation.

A further theme which we have stressed in discussing the engineering case-study was the way in which management chose to rely upon the craft ethos; in other words, it was prepared to accept the relative autonomy of the craftsman. To some degree, such dependence upon the worker was also traditional in the insurance company. In that instance it was declining, although in brewing there were some signs of an increase. However, if we combine these elements we find that there is no consistent relationship between the degree of joint regulation and the degree of worker autonomy. Furthermore, the ways in which managements in the two cases of traditional worker autonomy sought to ensure a degree of worker compliance with management goals were markedly different. In one there was stress upon long-term financial rewards, while in the other financial rewards were very immediate, in the form of piecework. The former involved a rather more bureaucratic structure of general labour regulation, and this form of career structure has been widely commented upon in the labour-process literature. However, the significance of piecework as a form of labour control has recently been largely ignored, despite its continuing significance (Gourlay 1986). Only Burawoy has discussed it, but in terms of a diversionary ploy (1979), in contrast to the much longer tradition within industrial relations which stressed, particularly in the late 1960s and early 1970s, as we have found, that it provided a series of very real controls for workers.

Table 4 on areas of joint regulation includes a variety of features which have recently received remarkably little attention in the discussion of labour control. These include such issues as the selection, training, and allocation of labour and manning levels. These are of crucial importance to labour control. Furthermore, they have important implications not only for the welfare of workers, but also for the efficient utilization of labour, a crucial consideration for management. Again, these themes have been widely discussed in the industrial relations literature, and more recently in terms of management strategy in a period of high unemployment. But they have rarely entered into more formal discussions within the labour-process tradition. We have seen, most clearly in the brewing and insurance companies, that these issues are of major importance as management seeks to grapple with signals from the marketplace. They may raise, in a particularly acute form, the complex balance of common and conflicting interests of capital and labour, along with the question of the balance of power. Their significance may lie less in the precise forms of control than in terms of the degree of management and worker control. This is highlighted by the extent to which bargaining occurred over such matters, and to which management was induced to modify its initial proposals. Hence, a crucial question in labour control terms is precisely how far management can recruit, shed, and utilize labour as it wishes, even within the primary labour market sector to which all of our firms belonged.

In some cases, factors such as recruitment and training may assume particular significance for management as a means of easing problems of labour control and efficient utilization. They did not appear to be particularly significant in the companies studied here, although the brewery had a youth training scheme and temporary employment was being used as a probationary period in the chemical company. More important, we would suggest, was the simple fact that all four of our companies tended to pay above average rates for comparable jobs in their areas. This meant that they could recruit more easily, and it played some role in ensuring a degree of worker compliance (cf. Thurow 1975). Again, while this is a factor recognized within dual labour market theories, it has not been sufficiently incorporated into the mainstream labour-process debate.

This point leads on to three other considerations. First, a great deal of the labour-process literature has focused upon forms of control within the workplace. But not only may workers gain satisfaction from certain features of the work process (see e.g. Baldamus 1961; Fox 1971; Warr 1983), but in addition there are important pressures emanating from the broader society for workers to conform to the demands of the production process. In many respects the primary forms of control are relatively simple: an employer pays the going rate or better and seeks to be seen as acting fairly and responsibly. Other forms of control are basically compromises and accommodations with the demands of the workforce—a series of quid pro quos. Indeed, that is basically what the wage–effort bargain is. Certainly the partners to the labour contract do not enjoy an equality of power. But, given the extent to which the employer is dependent upon the co-operation and compliance of the workforce, he is generally ill-advised to 'punch his weight'. Structures of control therefore exist, but the primary forms of control, the factors that generally ensure a modicum of worker compliance, derive in large part from broader structural features of our society—the need for an income, the very ordinariness and routine nature of work, etc. (see e.g. Braverman 1974; Burawoy 1985). In this sense internal structures of control may be seen as relatively marginal.

Second, as we have stressed in looking at management rationales for introducing new technology, labour control is not the primary means by which employers achieve their goal of profitability. This depends upon many other factors. Accordingly, we have suggested, there is a need to locate labour issues within this broader context, and—particularly as far as task-level controls are concerned—to see them in relation to the broader system of production control. Too often writers have treated production control systems as if they were solely or primarily orientated towards labour. Recently, as was noted in Chapter 1, there has been a growing recognition of this weakness.

Third, a question arises as to the degree of coherence within management control strategies. In recent years there have, ironically, been quite different developments in ideas on this issue. Those who have been concerned with notions of the flexible labour force have suggested that there is a growing coherence in management strategy (e.g. Clutterbuck 1985; however, more recent writings within this tradition have become somewhat more cautious, e.g. Meager 1986), while others within the labour-process tradition have increasingly questioned the extent to which one can meaningfully talk about a coherent management strategy (see Chapter 1). The degree of coherence, it seems to us, is an important and interesting question which is to be answered empirically.

What does appear to be common, however, is some disjunction between patterns of task-based regulation and the more general structure of labour regulation. As we have noted, joint control features of the latter may act as an important constraint upon variations in task-based arrangements. But, in addition, the significance of particular features of 'labour control' may vary widely. For example, various techniques which were once seen as attracting employees and fostering corporate identity, such as on-site canteens, have now become largely taken for granted (Batstone 1981). The significance of formal grievance procedures often owes less to the desire of an employer to create an 'internal state' than to his fear of being accused of unfair dismissal (Brown 1981). But once broader structures of labour regulation exist, they may begin to constrain an employer's actions at the level of the task and utilization of labour.

Fourth and relatedly, a great deal of stress has been placed upon management strategy, and rather less upon its efficacy. Particularly in the case of techniques aimed at winning worker commitment, relatively little attention has been paid to the question of how far the extent to which workers are likely to 'comply' with such structures (cf. the discussion of Etzioni in Chapter 1). We can usefully highlight this, and a number of related points, by considering the nature of internal labour markets.

We have already noted the importance of union regulation in this respect. We suggested that the significance of such career structures and mobility options depended, among other things, upon the extent to which other options existed. Relatedly, it is necessary to look more closely at the particular features of any form of labour regulation. In a number of our case-studies persons of different grades worked closely together; there was therefore an overlap between the promotion route and the structure of authority. In other words, assuming a degree of commitment to promotion, a worker was unlikely to object to the level of supervision—for the greater the number of supervisors the greater the chances of promotion. This point has been developed at some length by Grint (1985) in his study of postal workers in the inter-war period.

The kinds of occupational hierarchies found in a number of our case-studies also indicate that authority is dissipated. Close relations will often be found, for example, between an operator and the person immediately in authority over him—the chargehand or junior chargehand. Such diffusion of authority is seen by some radical writers as an important means of obscuring the true source of authority. But even if this were the origin of complex structures, there is no guarantee that they in fact operate in this way. For precisely because those in authority work with their subordinates there may be very strong pressures upon them to co-operate with rank-and-file workers and thereby to manipulate or evade management instructions. This has been commonly noted in the case of foremen (e.g. Child and Partridge 1982). In addition, workers with some degree of authority are frequently key union activists—as we noted, for example, in our insurance study. While in some situations their position within the authority structure may shape the way in which they approach their union role, the latter may equally affect their behaviour as supervisors.

Often occupational hierarchies do not merely graduate authority relationships but also differentiate between workers in terms of skills and production-related responsibilities. Those higher up the hierarchy are deemed to be more experienced or competent and therefore undertake the more complex and responsible tasks. Again, it might be argued that this serves to obscure authority relations since they become legitimated in terms of competence: this does seem likely. But of possibly greater significance is that the occupational hierarchy reflects less a concern to control workers, than a desire to ensure competence in the execution of production-related tasks. Indeed, while we did not investigate in detail the origins of the internal labour markets in our various companies, our impression was that they reflected factors other than a concern with obscuring the acquisition of surplus value or a desire to divide and rule. Rather, the division of labour had been constructed in the belief that workers required particular qualifications and experience before they could be expected to carry out specific tasks competently. In other words, they reflected a degree of company specificity of skill. The second factor, however, was that, given this finely graded structure, developing rules on promotion routes was in part a response to worker demands and one which, in the main, was relatively inexpensive for management. A by-product of this, however, was that higher-graded workers may often have been effectively locked into the firm—not necessarily because of a commitment to the promotion route as such, but because comparable earnings could not be obtained elsewhere. In this sense, the significance of internal labour markets has a rather more negative connotation than is frequently assumed. It is less that workers become committed to management goals in their desire for promotion: rather, they cannot afford to move jobs. It is a constraint upon external mobility.

There are also significant variations in the nature of internal labour markets. In our engineering company, for example, most mobility opportunities were of a horizontal nature. We have also suggested that such moves were of considerable significance in the chemical company. In other cases, as in insurance, vertical mobility was more significant. The former required less commitment to management goals than the latter—it often seemed to be a means of escape rather than anything more positive. But even here there was scope for management discretion, in the sense that they could pick and choose between applicants for promotion or transfer. More generally, one can conceive of a wide variety of internal labour markets. At one extreme is the situation in which senior managers simply choose who should be promoted, taking little account of workers' wishes. In such cases, it may be very difficult for the chosen worker to refuse (this has been the classic situation within management grades—although, interestingly, many managers have in recent years refused promotion, particularly where this had required moving home). In this situation, management has a very high level of discretion. At the other extreme is the strict seniority ladder where years of service are almost the sole criterion for promotion. Traditionally this system operated in the steel industry and applied not to the plant as a whole, nor to a particular department, but to a specific furnace or mill. Management discretion was therefore negligible. It would seem to follow that the extent to which internal labour markets provide management with control will vary widely. Such control may often be more significant in terms of the costs to workers of dismissal than in terms of the gains to be achieved by enthusiastic commitment to management wishes.

One implication of our preceding discussion is that the extent to which internal labour markets constitute a positive form of control has often been grossly exaggerated, even in terms of the intentions of employers. Where they reflect an 'incorporating' strategy, their significance may be essentially negative, particularly for manual workers. The importance of internal labour markets often lies more in the existence of job-specific skills (limited though these may be in practice) and accommodating worker pressures (see e.g. Becker 1964; Williamson 1975; Osterman 1984). In addition, the efficacy of internal promotion opportunities as a form of control depends upon the values and orientations of the workers concerned. It is clear that aspirations for promotion vary greatly and that many workers attach little value to such opportunities, particularly if it requires exemplary behaviour on their part for a number of years (see e.g. Lockwood 1966). Furthermore, failure to achieve promotion may lead to considerable dissatisfaction on the part of workers: it is not unusual, for example, for this to be a stimulus to union activism. In other words, structures of promotion may easily become counter-productive.

Many of the points which we have raised concerning internal labour markets apply with equal force to other forms of labour regulation. Attempts at technical control, for example, may often increase dependence upon certain groups of workers, as in the case of the control room operator in chemicals. Moreover, to the extent that such tasks involve 'second order' monitoring and control—monitoring the performance of the machine—it may become even more difficult to control and supervise the worker. In any event, as we noted in Chapter 1, rarely do machines actually determine what a worker must do: the assembly line, for example, simply brings work to a work station and moves it on after a certain period of time. It does not guarantee that the worker undertakes the expected series of tasks within the specified time period. There are, of course, strong pressures upon the worker to conform to the sets of expectations which are implicit in the design of the technology, but they derive only in part from the technology. The same, we have seen, is true of work rules and supervision. Similarly, piecework embodies contradictions, as has been widely recognized (for other examples see Batstone 1984).

Finally, we can briefly touch upon another theme which has been a considerable topic of discussion in recent years. This concerns the extent to which employers, in the face of slack product and labour markets, have had increasing resort to secondary labour. In all four of our companies use had certainly been made of such labour in recent years. In the engineering and insurance companies this had been done to meet exceptional demands—unusual jobs or problems during relocation. In the chemical company, temporary workers were employed when demand unexpectedly increased after manning had been reduced and there was uncertainty over future market trends. In all of these instances, management's use of secondary labour was of an *ad hoc*, rather than a strategic, nature. In two cases, however, rather more long-term resort to secondary labour can be seen. In the insurance company catering had been subcontracted. In the brewery similar attempts were made, but in the face of concerted union opposition management was able to achieve only a small proportion of the total contract work it would have liked. Furthermore, the amount of work contracted remained less than it had been in the early 1970s, when much greater reliance had been placed on contract distribution. In the engineering company the amount of work contracted out had been substantially reduced. This picture of variable trends is consistent with the picture found in our survey (Batstone and Gourlay 1986). The typically *ad hoc* nature of resort to secondary labour also conforms with the picture emanating from a good deal of other work (e.g. Meager 1986). In none of our companies, however, did secondary labour account for more than 1 per cent of labour usage.

In this chapter we have touched upon a wide range of issues in seeking to relate our findings to a series of more general debates and themes. It should, of course, be stressed that there are limitations to how far and in what ways one can generalize from a small number of case-studies. However, the similarity of the findings reported here with those of our survey (Batstone and Gourlay 1986) and with other case-studies gives us somewhat greater confidence than we would otherwise have in seeking to make general statements. It is therefore useful to summarize our key findings once more. First, there are no unilinear trends in the way in which work organization and experience are shaped by the introduction of new technology. More than this, different elements of the task situation are frequently affected in different ways. Second,

rarely are labour considerations central to management
thinking about new technology, although the implications for
labour of management rationales are variable. Third, the
extent to which labour issues shaped management thinking
after the decision had been made to introduce new technology
was very variable. Often issues of work organization were
decided late in the day and through chance factors. In
addition, broader industrial relations considerations sometimes played an important role. Fourth, the degree of union
influence over these matters varied significantly. Fifth, union
influence over new technology reflected its more general role,
and this, in turn, was closely related to the degree of union
sophistication. In explaining the latter we highlighted the
interaction between union organization, membership attitudes,
and the nature of the work situation. Clearly, management
can affect all three of these factors, as has been widely
recognized. But, as a growing body of literature indicates, the
role of trade unions is also a significant and partially
independent force.

REFERENCES

Adams, R. J. (1975). *The Growth of White-Collar Unionism in Britain and Sweden*, Wisconsin: University of Wisconsin Press.

Anthony, P. D. (1977). *The Ideology of Work*, London: Tavistock.

Argyris, C. (1972). *The Applicability of Organizational Sociology*, Cambridge: Cambridge University Press.

Armstrong, P. (1986). 'Management control strategies and inter-professional competition', in D. Knights and H. Willmott (eds.), *Managing the Labour Process*, Aldershot: Gower.

Bain, G. S. (1970). *The Growth of White-Collar Unionism*, Oxford: Clarendon Press.

—— Coates, D., and Ellis, V. (1973). *Social Stratification and Trade Unionism*, London: Heinemann.

Baldamus, W. (1961). *Efficiency and Effort*, London: Tavistock.

Barras, R., and Swann, J. (1983). *The Adoption and Impact of Information Technology in the UK Insurance Industry*, London: Technical Change Centre.

Barron, I., and Curnow, R. (1979). *The Future with Microelectronics*, Milton Keynes: Open University Press.

Batstone, E. (1979). 'Systems of domination, accommodation and industrial democracy', in T. Burns *et al.* (eds.), *Work and Power*, London: Sage.

—— (1981). 'Food, welfare and labour control', duplicated paper, Nuffield College, Oxford.

—— (1984). *Working Order*, Oxford: Blackwell.

—— Boraston, I., and Frenkel, S. (1975). 'Orientation to work and the negotiation of meaning', duplicated paper, Industrial Relations Research Unit, University of Warwick.

—— —— —— (1977). *Shop Stewards in Action*, Oxford: Blackwell.

—— Ferner, S., and Terry, M. (1984). *Consent and Efficiency*, Oxford: Blackwell.

—— and Gourlay, S. (1986). *Unions, Unemployment and Innovation*, Oxford: Blackwell.

Becker, G. S. (1964). *Human Capital*, New York: Columbia University Press.

Bell, D. (1974). *The Coming of Post-Industrial Society*, London: Heinemann.

Blackburn, R. M., and Mann, M. (1979). *The Working Class in the Labour Market*, London: Macmillan.

Blackburn, R. M. and Prandy, K. (1965). 'White-collar unionism: a conceptual framework', *British Journal of Sociology*, 16.

Blauner, R. (1964). *Alienation and Freedom*, Chicago: Chicago University Press.

Brady, T. (1984). *New Technology and Skills in British Industry*, Falmer, Brighton: Science Policy Research Unit, University of Sussex.

Braverman, H. (1974). *Labor and Monopoly Capital*, New York: Monthly Review Press.

Brown, W. (1973). *Piecework Bargaining*, London: Heinemann.

—— (ed.) (1981). *The Changing Contours of British Industrial Relations*, Oxford: Blackwell.

Burawoy, M. (1979). *Manufacturing Consent*, Chicago: Chicago University Press.

—— (1985). *The Politics of Production*, London: Verso.

Carchedi, G. (1977). *On the Economic Identification of Social Classes*, London: Routledge and Kegan Paul.

Chandler, A. (1962). *Structure and Strategy*, Cambridge, Mass.: MIT Press.

Child, J., and Partridge, B. (1982). *Lost Managers*, Cambridge: Cambridge University Press.

Clutterbuck, D. (ed.) (1985). *New Patterns of Work*, Aldershot: Gower.

Crompton, R. (1976). 'Approaches to the study of white-collar unionism', *Sociology*, 10.

—— (1979). 'Trade unionism and the insurance clerk', *Sociology*, 13.

—— and Jones, G. (1984). *White-Collar Proletariat*, London: Macmillan.

Dalton, M. (1959). *Men Who Manage*, New York: Wiley.

Davies, A. (1986). *Industrial Relations and New Technology*, London: Croom Helm.

Davis, L. E., and Taylor, J. C. (1976). 'Technology, organization and job structure', in R. Dubin (ed.), *Handbook of Work, Organization and Society*, Chicago: McNally.

Deaton, D. (1985). 'Management style and large-scale survey evidence', *Industrial Relations Journal*, 16.

Doeringer, P., and Piore, M. (1971). *Internal Labor Markets and Manpower Analysis*, Lexington, Mass.: Heath Lexington.

Dubois, P., Durand, C., Chave, D., and Le Maitre, G. (1976). *L'Autonomie ouvrière dans les industries des série*, Paris: Groupe de Sociologie du Travail.

Duhm, R., and Muchenberger, U. (1983). 'Computerization and control strategies at plant level', *Policy Studies*, 3.

Edwards, C., and Heery, E. (1985). 'Formality and informality in the Coal Board's national incentive scheme', *British Journal of Industrial Relations*, 16.

Edwards, P., and Scullion, H. (1982). *The Social Organization of Industrial Conflict*, Oxford: Blackwell.

Edwards, R. (1979). *Contested Terrain*, London: Heinemann.

Elbaum, B. (1984). 'The making and shaping of job and pay structures in the iron and steel industry', in P. Osterman (ed.), *Internal Labor Markets*, Cambridge, Mass.: MIT Press.

Etzioni, A. (1961). *A Comparative Analysis of Complex Organizations*, New York: Free Press.

Evans, C. (1979). *The Mighty Micro*, London: Gollancz.

Fox, A. (1971). *The Sociology of Work in Industry*, London: Collier-Macmillan.

—— (1974). *Beyond Contract*, London: Faber.

Friedman, A. (1977). *Industry and Labour*, London: Macmillan.

Gallie, D. (1978). *In Search of the New Working Class*, Cambridge: Cambridge University Press.

Glenn, E. N., and Feldberg, R. L. (1979). 'Proletarianizing clerical work', in A. Zimbalist (ed.), *Case Studies in the Labor Process*, New York: Monthly Review Press.

Goodrich, C. L. (1920). *The Frontier of Control*, new edition 1975, London: Pluto.

Gospel, H. (1983). 'Managerial structures and strategies', in H. Gospel and C. Littler (eds.), *Managerial Strategies and Industrial Relations*, London: Heinemann.

Gourlay, S. (1986). 'The ambiguity of control: a case from small batch engineering', paper presented to UMIST–Aston Labour Process Conference.

Grint, K. (1985). 'Bureaucracy and Democracy: The Quest for Industrial Control in the Postal Business between the Wars', University of Oxford, D.Phil. thesis.

Heritage, J. (1980). 'Class situation, white-collar unionization and the "double proletarianization" thesis: a comment', *Sociology*, 14.

Hickson, D. J., Pugh, D. S., and Pheysey, D. C. (1969). 'Operations technology and organization structure', *Administrative Science Quarterly*, 14.

Hill, S. (1981). *Competition and Control at Work*, London: Heinemann.

Hilton, J., et al. (1935). *Are Trade Unions Obstructive?*, London: Gollancz.

Hirshhorn, L. (1984). *Beyond Mechanization*, Cambridge, Mass.: MIT Press.

Hodson, R. (1983). *Workers' Earnings and Corporate Economic Structure*, New York: Academic Press.

Hyman, R. (1979). 'The politics of workplace trade-unionism', *Capital and Class*, 8.

Hyman, R. and Price, R. (eds.) (1983). *The New Working Class? White-Collar Workers and Their Organizations*, London: Macmillan.

Jacoby, S. M. (1984). 'The development of internal labour markets in American manufacturing firms', in P. Osterman (ed.), *Internal Labor Markets*, Cambridge, Mass.: MIT Press.

Jaques, E. (1967). *Equitable Payment*, Harmondsworth: Penguin.

Jones, B. (1982). 'Destruction or redistribution of engineering skills? The case of numerical control', in S. Wood (ed.), *The Degradation of Work?*, London: Hutchinson.

de Kadt, M. (1979). 'Insurance: a clerical work factory', in A. Zimbalist (ed.), *Case Studies on the Labor Process*, New York: Monthly Review Press.

Kelly, J. (1985). 'Management's redesign of work: labour process, labour markets and product markets', in D. Knights, H. Willmott, and D. Collinson (eds.), *Job Redesign*, Aldershot: Gower.

Kerr, C. (1983). *The Future of Industrial Societies*, Cambridge, Mass.: Harvard University Press.

—— et al. (1973). *Industrialism and Industrial Man*, Harmondsworth: Penguin (1st edn. published by Harvard University Press, 1960).

Knights, D., Willmott, H., and Collinson, D. (eds.) (1985). *Job Redesign*, Aldershot: Gower.

Kochan, H., McKersie, R., and Cappeli, P. (1984). 'Strategic choice and industrial relations theory', *Industrial Relations*, 23.

Krieger, J. (1983). *Undermining Capitalism*, London: Pluto.

Kuhn, J. W. (1961). *Bargaining in Grievance Settlement*, Columbia: Columbia University Press.

Kusterer, K. C. (1978). *Know-How on the Job*, Boulder, Colo.: Westview Press.

Levie, H., and Moore, R. (eds.) (1984). *The Control of Frontiers*, Oxford: Ruskin College.

Littler, C. (1982). *The Development of the Labor Process in Capitalist Societies*, London: Heinemann.

Lockwood, D. (1958). *The Black-Coated Worker*, London: Allen and Unwin.

—— (1966). 'Sources of variation in working-class images of society', *Sociological Review*, 14.

Lupton, T. et al. (1979). 'Manufacturing system design in Europe', in C. Cooper and E. Mumford (eds.), *The Quality of Working Life in Western and Eastern Europe*, London: Associated Business Press.

McGregor, D. (1960). *The Human Side of Enterprise*, New York: McGraw-Hill.

Maitland, I. (1983). *The Causes of Industrial Disorder*, London: Routledge and Kegan Paul.

Mallet, S. (1975). *The New Working Class*, Nottingham: Spokesman.

Manpower Services Commission (1985). *The Impact of New Technology on Skills in Manufacturing and Services*, Sheffield: MSC.

Meager, N. (1986). 'Temporary workers', *Employment Gazette*, January.

Meissner, M. (1969). *Technology and the Worker*, San Francisco: Chandler.

Merkle, J. (1980). *Management and Ideology*, Berkeley: California University Press.

Mills, C. Wright (1951). *White Collar*, New York: Oxford University Press.

National Board for Process and Incomes (1967). *Payment by Results*, Report No. 65, Cmnd. 3627. London: HMSO.

Nelson, D. (1975). *Managers and Workers*, Madison, Wis.: University of Wisconsin Press.

Nichols, T., and Beynon, H. (1977). *Living with Capitalism*, London: Routledge and Kegan Paul.

Nisbet, R. (1971). 'The impact of technology on ethical decision-making', in J. Douglas (ed.), *The Technological Threat*, New York: Prentice Hall.

Nolan, P. (1983). 'The firm and labour market behaviour', in G. S. Bain (ed.), *Industrial Relations in Britain*, Oxford: Blackwell.

Northcott, J., and Rogers, P. (1982). *Microelectronics in Industry*, London: PSI.

Offe, C. (1976). *Industry and Inequality*, London: Arnold.

Ogden, S. R. (1982). 'Bargaining structure and the control of industrial relations', *British Journal of Industrial Relations*, 20.

Osterman, P. (ed.) (1984). *Internal Labor Markets*, Cambridge, Mass.: MIT Press.

Penn, R., and Scattergood, H. (1985). 'Deskilling or enskilling?', *British Journal of Sociology*, 36.

Perrow, C. (1970). *Organizational Analysis*, London: Tavistock.

Pettigrew, A. (1973). *The Politics of Organizational Decision-Making*, London: Tavistock.

Prandy, K., Stewart, A., and Blackburn, R. M. (1983). *White-Collar Unionism*, Cambridge: Cambridge University Press.

Price, R. (1983). 'Introduction', in R. Hyman and R. Price (eds.), *The New Working Class?*, London: Mamillan.

Pugh, D. S., and Hickson, D. J. (1976). *Organizational Structure in its Context*, London: Saxon House.

Rajan, A. (1984). *New Technology and Employment in Insurance, Banking and Building Societies*, Aldershot: Gower.

Rayton, D. (1972). *Shop-Floor Democracy in Action*, Nottingham: ICOM.

Reeves, T. K., and Woodward, J. (1970). 'The study of managerial control', in J. Woodward (ed.), *Industrial Organization: Behaviour and Control*, Oxford: Oxford University Press.

Rhodes, E., and Wield, D. (eds.) (1985). *Implementing New Technologies*, Oxford: Blackwell.

Rose, M., and Jones, R. (1985). 'Managerial strategy and trade union responses in work reorganization schemes at establishment level', in D. Knights *et al.* (eds.), *Job Redesign*, Aldershot: Gower.

Rubery, J. (1978). 'Structured labour markets, worker organization and low pay', *Cambridge Journal of Economics*, 2.

Sayles, R. (1958). *The Behavior of Industrial Work Groups*, New York: Wiley.

Schattsneider, E. E. (1960). *The Semi-Sovereign People*, New York: Hold, Rinehart, and Wilson.

Slichter, S. S., Healy, J. J., and Livernash, E. R. (1960). *The Impact of Collective Bargaining on Management*, Washington: Brookings Institute.

Social Science Research Council (1968). *Social Research on Automation*, London: Heinemann.

Sorge, A., *et al.* (1983). *Microelectronics and Manpower in Manufacturing*, Aldershot: Gower.

Stinchcombe, A. (1970). 'Bureaucratic and craft administration of production', in O. Grusky and G. A. Miller (eds.), *The Sociology of Organizations*, New York: Free Press.

Storey, J. (1986). 'The phoney war? New office technology: organization and control', in D. Knights and H. Willmott (eds.), *Managing the Labour Process*, Aldershot: Gower.

—— (1985). 'The means of management control', *Sociology*, 19.

Supple, B. (1970). *The Royal Exchange Assurance*, Cambridge: Cambridge University Press.

Terry, M. (1978). 'Shop stewards: the emergence of a lay elite?', discussion paper, Industrial Relations Research Unit, University of Warwick.

Thompson, P. (1983). *The Nature of Work*, London: Macmillan.

Thurow, L. (1975). *Generating Inequality*, New York: Basic Books.

Touraine, A. (1971). *The Post-Industrial Society*, New York: Random House.

Turner, A., and Lawrence, P. R. (1965). *Industrial Jobs and the Worker*, Harvard: Harvard University Press.

Turner, H. A. (1962). *Trade Union Growth, Structure and Policy*, London: Allen and Unwin.

van Houten, D. (1980). Review of *Contested Terrain* by R. Edwards in *Economic and Industrial Democracy*, 1.

Walton, R. E., and McKersie, R. B. (1965). *A Behavioral Theory of Labor Negotiations*, New York: McGraw-Hill.

Warr, P. (1983). 'Job loss, unemployment and psychological well-being', in E. van de Vliert and V. Allen (eds.), *Role Transitions*, New York: Plenum.

Wedderburn, D., and Crompton, R. (1972). *Workers' Attitudes and Technology*, Cambridge: Cambridge University Press.

Wilkinson, B. (1983). *The Shop-Floor Politics of New Technology*, London: Heinemann.

Williamson, E. O. (1975). *Markets and Hierarchies*, New York: Free Press.

Willman, P. (1980). 'Leadership and trade union principles', *Industrial Relations Journal*, 11.

Winkler, J. (1974). 'The ghost at the bargaining table', *British Journal of Industrial Relations*, 12.

Woodward, J. (1965). *Industrial Organization: Theory and Practice*, Oxford: Oxford University Press.

—— (ed.) (1970). *Industrial Organization: Behaviour and Control*, Oxford: Oxford University Press.

Yanouzas, J. (1964). 'A comparative study of work organization and supervisory behavior', *Human Organization*, 26.

Zeitlin, J. (1983). 'Social theory and the history of work', *Social History*, 8.

INDEX